Fundamental Concepts and Computations in Chemical Engineering

FUNDAMENTAL CONCEPTS AND COMPUTATIONS IN CHEMICAL ENGINEERING

Vivek Utgikar

Boston • Columbus • Indianapolis • New York • San Francisco • Amsterdam • Cape Town
Dubai • London • Madrid • Milan • Munich • Paris • Montreal • Toronto • Delhi • Mexico City
São Paulo • Sydney • Hong Kong • Seoul • Singapore • Taipei • Tokyo

For information about buying this title in bulk quantities, or for special sales opportunities (which may include electronic versions; custom cover designs; and content particular to your business, training goals, marketing focus, or branding interests), please contact our corporate sales department at corpsales@pearsoned.com or (800) 382-3419.

For government sales inquiries, please contact governmentsales@pearsoned.com.

For questions about sales outside the U.S., please contact intlcs@pearson.com.

Visit us on the Web: informit.com/aw

Library of Congress Control Number: 2016948715

ISBN-13: 978-0-13-459394-4
ISBN-10: 0-13-459394-4

Text printed in the United States on recycled paper.
1 16

This book is dedicated to the late Sharayu Prabhakar Utgikar and Prabhakar Vasant Utgikar.

CONTENTS

PREFACE

The first semester of the undergraduate chemical engineering program at the University of Idaho features two courses in chemical engineering: *Introduction to Chemical Engineering* and *Computations in Chemical Engineering*. The first of these courses provides the students a comprehensive exposure to the nature of the chemical engineering field as well as to the wide variety of career opportunities available to them upon completion of their chemical engineering education. The students get a glimpse into the types of activities that comprise the responsibilities of different positions in the chemical engineering field. The second course gives them a flavor of the types of engineering calculations they will be expected to deal with during both their studies and their professional careers as chemical engineers. For the students, the value of these courses comes from understanding their possible career paths, discovering their abilities and competencies, and overall being in a better position to make an informed decision regarding their career choice early in their college education. The students also get to interact with the chemical engineering faculty and know them from the very first semester of their college education. The faculty, too, get to know and develop relationships with the students practically from their first day of college. These courses help the faculty assess the interest and aptitude of an individual to succeed as a chemical engineer, identify and assist individuals needing extra attention, as well as encourage, nurture, and mentor the truly gifted ones. Such introductory courses having these objectives are becoming a norm for the chemical engineering programs leading to a baccalaureate degree.

This book has grown out of the need for a text that accomplishes the objectives of these introductory courses and lays the foundation for students' success not only in their studies but also in their careers.

The motivation was to create a book that gives freshman chemical engineering students an excellent idea about what it takes to get a baccalaureate degree in chemical engineering, the nature and scope of the industries where they most likely will be employed, typical responsibilities in various positions, and possible computations they would perform in their jobs as chemical engineers.

Who Should Read This Book

This book has been developed for freshman chemical engineering students entering the first year of the undergraduate degree program. These students, with a few exceptions, are high school graduates who have a basic knowledge of mathematics and science at varying levels but no exposure to engineering concepts. The book assumes no specific preparation on the part of students other than graduation from high school with basic science (physics, chemistry) and mathematics knowledge. It is also assumed that a typical student has rudimentary computer skills, including email and basic operations (launching programs, editing, saving documents, etc.), with a software program. However, no advanced skills, such as data manipulation, are needed.

The book also serves as a handy, quick reference for chemical engineering fundamentals as well as information on chemical industry for anyone engaged in chemical engineering activities, including educators and industry professionals.

How This Book Is Organized

The book is structured to provide to students, through the first three chapters, an introduction to the chemical engineering profession, chemical and allied industries, and their progression through a typical four-year undergraduate chemical engineering curriculum. The remaining chapters deal with the computational problems in chemical engineering, arranged closely according to the chronological order of subjects students will encounter in the undergraduate curriculum.

Chapter 1, "The Chemical Engineering Profession," presents a brief introduction to the engineering field and the position of chemical engineering in the broader engineering profession. The role and nature of typical job functions of a chemical engineer in different types of jobs are also described.

Chapter 2, "Chemical and Allied Industries," focuses on the significance of the chemical and allied industries in the nation's economy, with an exposure to the largest chemical companies and the chemical products. Chapter 3, "Making of a Chemical Engineer," outlines a typical chemical engineering curriculum with brief descriptions of advanced undergraduate chemical engineering courses, the engineering science courses that prepare students for these advanced courses, and the science and mathematics courses that provide the foundation for studying engineering. The role of humanities and social science courses is also described.

The importance of computations and use of computational tools in chemical engineering is presented in Chapter 4, "Introduction to Computations in Chemical Engineering." The classification of problems on the basis of their mathematical nature is also described in this chapter. Chapters 5 through 9 deal with typical problems in chemical engineering, with each chapter dealing with one specific area. Chapter 5, "Computations in Fluid Flow," describes the fundamental fluid flow phenomena and presents associated computational problems in practical systems. Chapter 6, "Material Balance Computations," discusses the basic principles of material balance computations with example problems, while the concepts of energy balance are covered in Chapter 7, "Energy Balance Computations." Chapter 8, "Computations in Chemical Engineering Thermodynamics," and Chapter 9, "Computations in Chemical Engineering Kinetics," discuss fundamental principles of chemical engineering thermodynamics and kinetics, respectively, with selected problems in different topical areas.

Each chapter is arranged to provide a context to the type of problems in the particular field followed by a discussion of essential theoretical fundamentals. Representative example problems are presented and their solutions discussed in detail. Alternative solution techniques for most of the problems are demonstrated using two different, distinct software tools—a spreadsheet program (Excel) and Mathcad. Exercise problems are included at the end of each chapter to provide students an opportunity to practice and gain mastery over the solution techniques. Many of the example as well as practice problems in later chapters are linked to the problems in earlier chapters to emphasize the integrated nature of the practical systems and problems. A brief introduction to the world of software, including commercial process simulation software, is presented in the appendixes, primarily to make the students aware of the various alternative, powerful computational tools that are available to perform complex calculations on a very large scale.

A typical student enrolled in the introductory courses is an incoming freshman student, a recent high school graduate, who is also invariably concurrently enrolled in the first college chemistry and calculus courses. This creates a student body that exhibits a wide range of familiarity with basic concepts in mathematics and chemistry, depending on the rigor of their preparation in high school. This wide discrepancy in the student backgrounds was simultaneously a challenge and an opportunity for innovative thinking in creating the text. The material presented in the book accounts for the variance in student preparation and seeks to provide enough background for those who have not been exposed to the relevant topics in chemistry and calculus and yet to avoid making it too basic for those who have had such exposure. In light of the student preparation (or lack thereof) in calculus courses, the book has also steered clear of in-depth discussion of differential equations while presenting clear solution techniques for them in easily understandable language. The book is designed to offer an instructor maximum flexibility to explore and delve into topics at any depth appropriate for the class.

Graduation is the most significant event for high school students all over the world. For many, it is a finish-line marking the end of the "student phase" of their lives. It signifies for them the conclusion of their formal education and their readiness to enter the "real world." For many others, however, it is merely an important milestone marking the conclusion of one and beginning of another educational journey, this time in a college or a university. It is my hope and expectation that this book will serve as an illuminating guide for those who are choosing the path of chemical engineering for this journey.

Register your copy of *Fundamental Concepts and Computations in Chemical Engineering* at informit.com for convenient access to downloads, updates, and corrections as they become available. To start the registration process, go to informit.com/register and log in or create an account. Enter the product ISBN (9780134593944) and click Submit. Once the process is complete, you will find any available bonus content under "Registered Products."

ACKNOWLEDGMENTS

It has taken nearly three years from the time the idea for the book came to me, first to develop the outline of the book and then to actually write the book. Such an endeavor would not have been possible without a little bit (a lot, really) of help from numerous friends, colleagues, and many others. It is my pleasure to acknowledge my debt to these individuals who have helped me, directly or indirectly, to turn the idea into reality.

I would like to start by thanking all my teachers at the Institute of Chemical Technology (ICT), Mumbai, India (Bombay University Department of Chemical Technology – BUDCT or simply UDCT in its previous incarnation), and at the University of Cincinnati in Ohio, who provided the illumination for my own educational journey in chemical engineering. I am especially grateful to Professor J. B. Joshi, Mumbai, and Professor Rakesh Govind, Cincinnati, with whom I had the good fortune of conducting my graduate studies. I would also like to thank Professors Wudneh Admassu and Roger Korus, for believing in me enough to offer an opportunity to join the faculty of the University of Idaho. I am greatly indebted to Professor Richard Jacobsen, former Dean of Engineering at the University of Idaho and Idaho State University, who provided invaluable input to make this book far better than its initial form.

A very special mention must be made of Dr. David MacPherson at the University of Idaho. The *Computations* course at the University of Idaho typically has enough enrollment for three sections every fall, and it is entirely due to Dr. MacPherson that the university is able to serve all the students well. His desire and tireless, selfless efforts to help the students learn and succeed are an inspiration. I am extremely lucky to have such a delightful person as Dr. MacPherson to work with. I would also like to acknowledge and thank all the teaching assistants, especially Zachary Beaman, Michael Cron, Megan Dempsey, and Adam Spencer, for their dedication and invaluable help in teaching the course and mentoring the students.

I am grateful to Laura Lewin, executive editor at Prentice Hall, who gave me the opportunity to publish this book. I would like to thank the team from Prentice Hall. Michael Thurston, as the developmental editor, helped to review all chapters and gave me valuable suggestions on the content presentation. Olivia Basegio helped coordinate with the team at Prentice Hall. Kathleen Karcher was instrumental in securing permission for material used in the text. Special thanks are due to Carol Lallier, who has done an amazing job of copyediting, painstakingly going over every word to bring clarity to the presentation. I would also like to thank Susie Foresman and Julie Nahil, for managing the production of the book.

The technical reviewers of the book have been thorough and diligent, providing feedback, identifying errors, and suggesting additional content to improve the book. I would like to express my sincere appreciation and heartfelt thanks to Patrick Cirino, Supathorn Phongikaroon, and Wudneh Admassu for taking on the onerous task of reviewing the book and ensuring the accuracy of the content as well as improving its presentation.

I would like to thank my wife and my sons for their support, encouragement, and understanding while I worked on the book.

Lastly and most importantly, I am thankful to all the students I have had the pleasure of teaching, especially in the two introductory courses in chemical engineering. It has been an incredibly rewarding experience for me to interact with, teach, and in turn learn from each and every one of them who has aspired to become a chemical engineer.

ABOUT THE AUTHOR

Dr. Vivek Utgikar is a professor of chemical engineering in the Department of Chemical and Materials Engineering and the Associate Dean of Research and Graduate Education for the College of Engineering of the University of Idaho. He has also served as the director of the nuclear engineering program of the University of Idaho. Dr. Utgikar's teaching portfolio includes a broad range of chemical and nuclear engineering courses such as transport phenomena, kinetics, thermodynamics, electrochemical engineering, hydrogen, and spent nuclear fuel disposition/management. His research interests include energy systems, nuclear fuel cycle processes, modeling of multiphase systems, and bioremediation. He was a National Research Council Associate at the National Risk Management Research Laboratory of the U.S. Environmental Protection Agency in Cincinnati, Ohio, prior to joining the University of Idaho. Dr. Utgikar is a registered professional engineer with process development, design, and engineering experience in chemical industry, and he holds a Ph.D. in chemical engineering from the University of Cincinnati. His other degrees include bachelor's and master's degrees in chemical engineering from the Mumbai University, India.

CHAPTER 1

The Chemical Engineering Profession

Engineering is the art of organizing forces of technological change.

—Gordon Stanley Brown[1]

Engineering, with its attractive salary and continued projected demand growth, is consistently ranked among the most desirable professions in various surveys and reports, such as the ones published by the *U.S. News & World Report*. Yet, only a little more than half of the students entering the engineering programs as freshman actually end up with a degree in engineering [1]. One of the reasons contributing to this student attrition is that a typical high school senior/college freshman entering an engineering program has only a limited understanding of the engineering profession, and this lack of understanding is compounded by the inability of engineers to answer the simple question, *What do engineers really do?* [2]. Answers often heard include the following:

- "An engineer solves problems."
- "An engineer builds and creates machines, processes, structures, and so on."
- "An engineer makes things."

All of these answers are true and yet add little to the student's conception of the engineering profession. In particular, such answers do not help distinguish an engineer from a scientist, a confusion that is only enhanced by the fact that almost all engineering degrees are designated as bachelor of science. If an individual is to succeed in an engineering course and profession,

---

1. Former Dean of Engineering at the Massachusetts Institute of Technology; an accomplished teacher, researcher, and administrator who influenced engineering education post World War II. Quotation source: Scalzo et al, *Database Benchmarking: Practical Methods for Oracle and SQL Server*, Rampant Techpress, Kittrell, NC, 2006.

he/she must understand the basic nature of engineering [3]. This is particularly true of chemical engineering, the most recent of the four main engineering disciplines. This chapter attempts to explain the chemical engineering profession, starting with a brief discussion of what the profession of engineering means and what roles engineers—the individuals practicing the engineering profession—play.

1.1 Engineering and Engineers

Engineering has been considered as the practical application of science for over two centuries. ABET, Inc., formerly the Accreditation Board of Engineering and Technology (www.abet.org), the organization that evaluates and accredits college and university programs in applied science, computing, engineering, and technology in the United States and many other countries, defines engineering as follows:

> Engineering is the profession in which a knowledge of mathematical and natural sciences, gained by study, experience and practice, is applied with judgment to develop ways to utilize economically the materials and forces of nature for the benefit of mankind.

This is an all-encompassing definition that makes it clear that engineering is based on science and harnesses resources for the benefit of mankind. Any activity practiced today as engineering would certainly be covered by this definition, and yet, the definition suffers from the disadvantage of being too general. Practically any commercial venture can be construed as an engineering activity according to this definition. Further, this definition does not help distinguish between science and engineering. Landis [4] has compiled 21 different definitions of engineering by various individuals over the last two centuries, and the recurring theme in these definitions is that *engineering is the application of science for the benefit of humanity*. As with the ABET definition, any engineering activity practiced today would conform to any one of these definitions, and yet, the definition would not be exclusive to engineering activities.

Obviously, defining engineering is not an easy task (we would have already had a more than adequate definition if it were so), and it would be better to describe characteristic attributes of engineering before attempting to develop a new definition for it:

- *An engineering activity has an economic impact associated with it:* An organization engages in engineering activity in order to derive immediate commercial or other benefit from it. The commercial benefit is primarily through generating revenue and profit through the manufacture of products, provision of service, supply of energy, and so on. The commercial benefit may also take the form of cost-aversion, such as avoiding penalties and taxes through the treatment of waste streams prior to releasing them in the environment. Certain engineering activities, particularly those conducted by governmental or quasi-governmental entities, may not have immediate or direct commercial benefit. For example, public works such as bridges, roads and dams, or weapons development and other military engineering activities may not generate revenue/profits but have implicit economic value to the society and the nation. These activities build infrastructure that makes economic growth possible and protect the people from external dangers that, among other impacts, threaten the economy of the nation. Similarly, engineering clean-up of polluted waterways and air and soils also has positive economic impact on the society through aversion of healthcare spending and other costs and increasing the resource availability for economic activities.

- *An engineering activity is conducted for the benefit of society at large:* Engineering activities are undertaken to satisfy the demand for products, services, and energy by the society. In other words, the public at large is the direct beneficiary of engineering activities, be those in the form of power plants, automobiles, machines, or clean environment. Some of these activities, particularly public works activities, benefit each and every member of the society. Other activities that are purely commercial in nature are targeted at the customers of the products and services of such activities. For example, an automobile manufacturer produces cars to satisfy the market demand for them. The customer for the product is any member of the society, any individual who has the need and resources to afford the product. This aspect of engineering activities contrasts with scientific activities, which while adding to our knowledge, are of direct benefit only to a more specialized group of people, namely other scientists, engineers, and technical professionals in that specific field.
- *Engineering involves application of scientific knowledge:* As the ABET definition states, engineering utilizes science for practical purposes. Science is essentially a field of discovery where fundamental understanding of processes and phenomena is the objective. Scientists seek to produce knowledge that enables us to explain observed

phenomena and elucidate underlying laws of nature. Engineers apply this knowledge to create products, processes, and services to improve the quality of life of people. A scientist is typically driven by the desire to derive a global understanding of the phenomena; in other words, the knowledge is not complete until each and every factor affecting the phenomena is identified and a theoretical explanation proposed and validated for every observed effect. Such knowledge is obviously beneficial to the engineer; however, an engineering activity can proceed without necessarily having the complete theoretical scientific understanding and framework for the process. As it is often said, *the steam engine was developed before the science (thermodynamics) behind it was fully understood*.

- *Engineering activities are conducted on a large scale:* A key significant characteristic that distinguishes engineering from science is the scale of activities. Science involves discovery of new products, principles, and processes in a laboratory or research setting. Engineering involves taking this novel discovery and scaling up to make it accessible and affordable to the entire society. A chemist may synthesize a new chemical with promising properties in the laboratory in milligram quantities, starting with high-purity substances. An engineer develops, on the basis of the reaction discovered in the laboratory, a commercial process producing the same chemical in bulk quantities (kilograms or tons, a scale-up factor of *a million or more*) starting with the cheapest substances possible. Scientists discovered radioactivity and nuclear fission; engineers utilized this knowledge to build nuclear power plants for electricity generation. The key challenge in engineering is to ensure that the outputs of industrial commercial ventures have the same properties (including purity) and functionality as the laboratory products while minimizing the energy and resource consumptions, and realizing economic benefit.

It should be noted that scale can refer to either the size or the number of the products of the engineering activity. Figure 1.1 shows the massive Hoover Dam on the Colorado River in the United States, a structure that required more than 5 million barrels of cement. Weighing in excess of 6.6 million tons, the Hoover Dam was the largest dam of its time when completed (www.usbr.gov/lc/hooverdam/faqs/damfaqs.html).

Not every product of engineering activities has such large dimensions. Figure 1.2(a) shows a 300-mm diameter silicon wafer containing a large

Figure 1.1 The Hoover Dam on Colorado River.

Source: Photo courtesy of the U.S. Bureau of Reclamation, www.usbr.gov/lc/hooverdam/images/D001a.jpg.

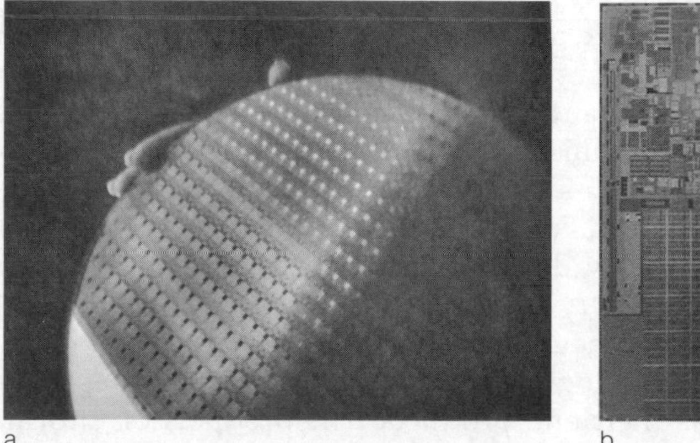

a.

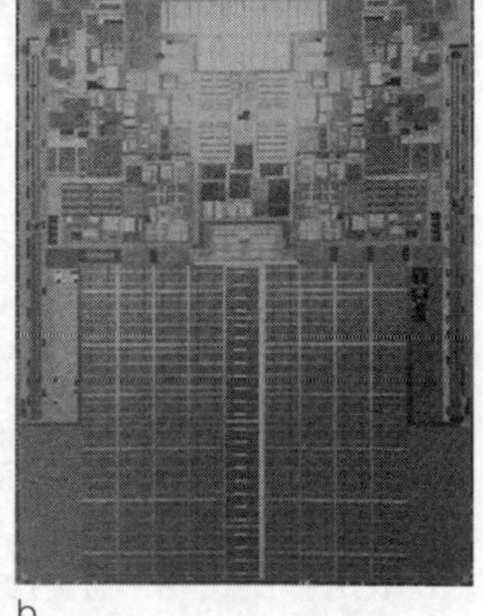

b.

Figure 1.2 A silicon wafer with microchips—(a) wafer, (b) a single microchip.

Source: Shackelford, J. F., *Introduction to Materials Science for Engineers*, Seventh Edition, Prentice Hall, Upper Saddle River, New Jersey, 2009.

number of microchips, each containing an integrated circuit similar to one shown in Figure 1.2(b) with the interconnections having a width of only a few nanometers [5]. Figures 1.1 and 1.2 both show manifestations of remarkable engineering endeavors.

These characteristics point out *engineering as the activity that makes science a reality for the common man*. It transforms the knowledge from the realm of a select few to an entity that can serve a consumer for betterment of the quality of life. The basic laws of mathematics, physics, and chemistry are harnessed through engineering to manufacture automobiles, planes, and other machines; build dams, highways, and other structures; construct power plants to supply electricity and refineries to provide gasoline; and countless such endeavors. A consumer, the beneficiary of these undertakings, does not need to have even a rudimentary understanding of the fundamental scientific laws to enjoy the resultant product or service, thanks to engineering.

An alternative way of expressing the same idea is to say that *engineering is the process of transforming science into technology*. This definition introduces an additional concept—that of *technology*. Technology is variously defined as (1) the application of science for practical purposes, (2) a branch of knowledge dealing with engineering or applied science as a machine, and (3) a piece of equipment developed from scientific knowledge. The first definition makes technology virtually indistinguishable from engineering, and the second definition lacks clarity. The third definition is the closest to the meaning intended in this book, as elaborated in the discussion that follows.

Science comprises the natural laws that govern the behavior and interaction of inanimate and animate objects. Technology is the manifestation or implementation of this knowledge in the form of a manufacturing, treatment, or other process that yields a machine, an object, or a service that is used by a consumer. For example, an automobile is the product of a manufacturing technology developed on the basis of laws of physics. Similarly, principles of chemistry are harnessed in another type of manufacturing technology to obtain a chemical product, such as sulfuric acid.

Not all technologies yield tangible objects such as sulfuric acid or an automobile as their end products. The products of information technology, based on mathematics and computer science, are frequently services in the form of software packages and programs used by consumers. Technology is essentially science-based process (or technique) that yields an object or a service that is used by the consumer, who need not have any understanding of the fundamental scientific principles on which it is based.

Once science is understood to be the foundation of the structure that is technology, engineering can be easily understood as the process of developing and building the structure. Engineering thus differs from both science and technology in being an action rather than a concept or an object. The initial starting point of engineering activities is science, and the end result is technology. Science and technology can be viewed as the initial and final states, and engineering, the process of traversing the path between them. Engineering also includes the actions that result in improvement of technology. This dynamic nature of engineering distinguishes it from both science and technology.

With this concept of engineering, an engineer can be conveniently defined as an individual engaged in the practice of engineering. An engineer must possess the scientific knowledge, but the overarching goal of the engineer is to apply this knowledge to create useful things—technology—for everyone. Discovering new knowledge is not an engineer's principal function, although some engineers may indeed add to the knowledge base while devising practical applications of science.

This scheme of things also allows us to define the role of a *technologist* with greater clarity. A technologist is the individual who has the responsibility to operate and ensure that the technology functions as designed and intended.[2] The technologist must understand how the technology operates; however, a technologist is not required to know *how* the technology was developed. A technologist may use his/her experience and empirical knowledge to effect improvements in the technology, but those activities would not qualify as engineering activities in the rigorous sense of engineering.

In the past, it was possible for an individual to acquire the necessary knowledge to practice the engineering profession through experience; however, at the present time a degree from an ABET-accredited program is indispensable for one to be qualified and licensed as a professional engineer.[3] These engineering programs feature rigorous science, engineering science, and engineering courses to prepare the individual for an engineering career.

Many universities also offer degree programs in engineering technology that prepare individuals for a career as a technologist. These programs are typically characterized by lower mathematical and scientific rigor and absence of the design component as compared to the engineering programs [3]. The emphasis is on understanding the operation of the process and maintenance of machinery. The actual operation of the machinery and the process is done by *technicians* skilled in the particular trade. The technical skills and knowledge necessary for such duty are

2. Despite the nature of responsibilities, these positions are most often termed as engineering positions.

3. Each state in the United States has its own board of licensed engineers that administers the necessary examinations to certify an individual as a professional engineer who can then legally engage in engineering practice.

typically acquired in a vocational school that offers an associate's degree or a certificate course or are learned on the job. The associate's degree or certificate programs are of shorter duration and feature less rigorous science and mathematics courses.

Based on the nature of engineering practice, the engineering field is divided into several disciplines. These disciplines are described briefly in the next section.

1.2 Engineering Disciplines

The evolution of engineering is essentially a history of civilization. As early humans transitioned from a nomadic existence into settlements, there arose a need for permanent structures and systems for handling water and wastewater demands. Civil engineering developed in response to these needs, and even today one can marvel at the systems developed by engineers of ancient civilizations in Egypt, Mesopotamia, Indus Valley, and Ancient Rome. Figure 1.3 shows Pont du Gard—a bridge over the river Gard in France built by the Roman engineers nearly 2000 years ago [6]. The third tier of this impressive structure, 160 feet above the river, is an aqueduct supplying water to the cities [6, 7].

Figure 1.4 shows the ruins of a sophisticated water reservoir from the city of Dholavira belonging to the Indus Valley Civilization dating several millennia further back. The cities and even smaller towns and villages of the Indus Valley Civilization feature a water and waste management system that can simply be described as outstanding [8].

The second driving force for engineering development was the need for accomplishing things beyond what was possible through simple manual labor, leading to simple machines such as the Archimedes screw (Figure 1.5), a simple yet elegant machine for pumping water and transferring material. This simple machine is capable of lifting the fluid, even if it contains a small amount of debris [9]. These machines progressively increased in complexity with time, driven in no small measure by the need for advanced weapons

Figure 1.3 A Roman bridge with the upper tier functioning as an aqueduct.
Source: Hanser, D. A., *Architecture of France*, Greenwood Press, Westport, Connecticut, 2006.

Figure 1.4 Ruins of a water reservoir in Dholavira.

Source: Archaeological Survey of India, http://asi.nic.in/images/exec_dholavira/pages/015.html.

Figure 1.5 A schematic representation of the Archimedes screw.

Source: Chondros, T. G., "Archimedes Life Works and Machines," *Mechanism and Machine Theory*, Vol. 45, No. 11, 2010, pp. 1766–1775.

and means of movement, leading to the discipline of mechanical engineering. This evolution has continued with new disciplines appearing over time in response to societal needs based on advances in sciences and new discoveries.

The American Society of Engineering Education (ASEE, www.asee.org), as a part of its activities, compiles statistical data on engineering graduates by disciplines, degree levels (bachelor, master, doctorate), demographics, and various other criteria. Figure 1.6 shows the breakdown by discipline of more than 106,000 bachelor's degrees in engineering awarded in 2014–15 [10]. Mechanical engineers form the largest group of engineering graduates, followed by civil, electrical, and chemical. These four disciplines are traditionally recognized as the big four of engineering, and any comprehensive school of engineering offers, at the minimum, degree programs in these four disciplines.

The emerging disciplines of engineering can also be identified from the figure. Computer science has the fourth-largest number of graduates, reflecting the explosive growth of information technology in recent times. However, it was not too long ago that most computer science programs

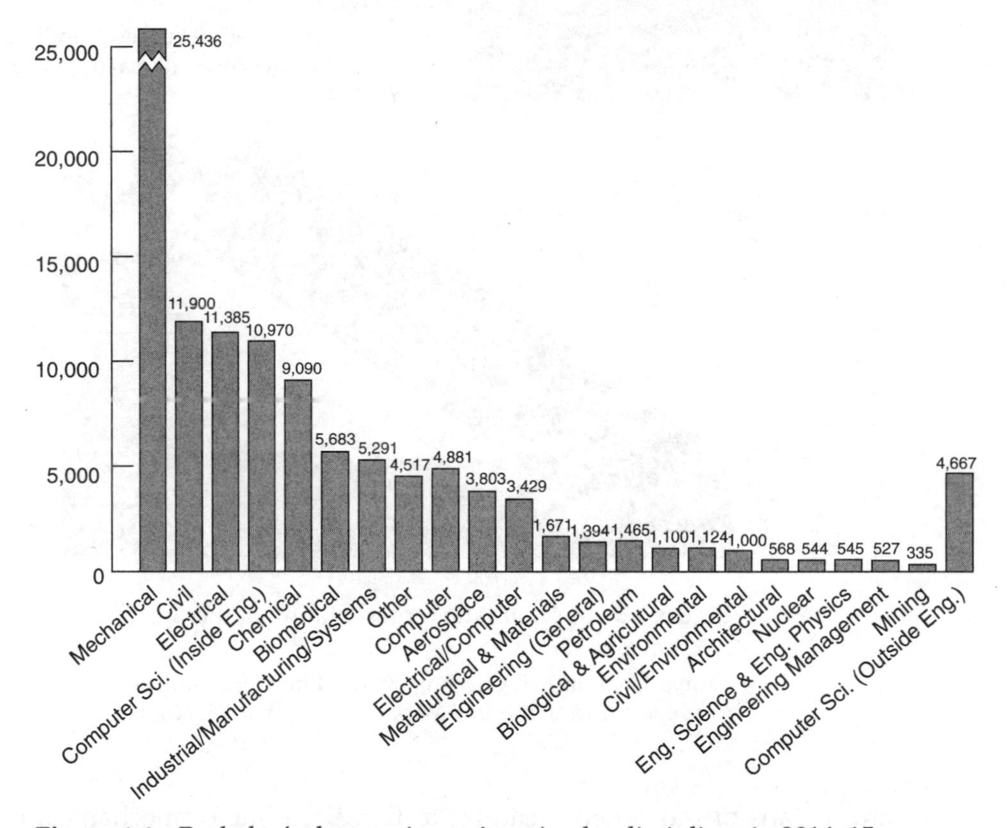

Figure 1.6 Bachelor's degrees in engineering by disciplines in 2014–15.
Source: Yoder, B. L., "Engineering by the Numbers," https://www.asee.org/papers-and-publications/publications/college-profiles/15EngineeringbytheNumbersPart1.pdf.

were housed in colleges of science, typically as a part of the mathematics department. The last two decades have seen these programs migrating to the college of engineering, with their own department that does not include the word *engineering* in its name. Computer science is distinct from computer engineering, which is often associated with electrical engineering programs. The other emerging engineering field is biomedical engineering, which has evolved out of advances in biotechnology and biomedical sciences.

The following briefly describes the big four engineering disciplines—mechanical, civil, electrical, and chemical:

- *Mechanical engineering*, as the name indicates, deals with development, design, and operation of all types of machines and systems for human convenience and comfort. This is the broadest of the engineering disciplines, with mechanical engineers serving fields ranging from machine design, manufacturing, energy, transportation machines and materials, materials handling, refrigeration and heating systems, maintenance and biomechanics, and many others. Mechanical engineers are a ubiquitous presence in all industries.
- *Civil engineering* is the earliest discipline of engineering dealing with the development, design, construction, and operation of the facilities and structures for the society. These facilities range from buildings, dams, highways, and canals to all types of systems and infrastructures for mass transit, water supply, waste disposition, and so on. Civil engineering can be further subdivided into specializations such as construction engineering, transportation engineering, geotechnical engineering, water resources engineering, and so on [4].
- *Electrical engineering* is the discipline dealing with machines and systems associated with electrical energy. Electrical engineers are concerned with the generation and transmission of electricity and with electrical machines. With electricity replacing all other types of energy for practically all consumer applications with the exception of automobiles and cooking, electrical engineers are as ubiquitous as mechanical engineers. Within electrical engineering, one can find specializations such as electronics, controls and power engineering, and so on.
- *Chemical engineering* as a discipline grew out of industrial chemistry and the need to produce large quantities of chemicals economically. Detailed discussion of chemical engineering is presented in the following section.

The identification of these four as distinct and major branches of engineering is not based on numbers alone but also on the fact that subjects studied during undergraduate studies for each are distinct enough to warrant a separate curriculum. There is an inevitable content overlap between disciplines, particularly between mechanical and civil engineering related to fluid and solid mechanics [11]. However, even for the overlapping content, the approach and treatment of topics and educational objectives generally differ for different disciplines. An individual obtaining a bachelor's degree in one of these disciplines will not be easily able to go on for graduate study in the other three disciplines without spending a substantial amount of time, perhaps equivalent to junior and senior years of a 4-year undergraduate degree program, completing prerequisite courses to address deficiencies. This is in contrast to the rest of the engineering disciplines seen in Figure 1.6.

With the possible exception of material and metallurgical engineering, an individual with a bachelor's degree in one of the big four disciplines can successfully go on to pursue graduate studies in many of the other engineering disciplines without needing substantial additional coursework to satisfy prerequisites. A civil, mechanical, or chemical engineer can go on to study and practice environmental engineering. A mechanical engineer can become an industrial, manufacturing, aerospace, or nuclear engineer. Similarly, a chemical engineer can pursue nuclear, biological/agricultural, or petroleum engineering graduate degrees. Many engineers, particularly over last two decades, have changed their fields to computer science at the graduate level. Many biomedical engineering programs operate only at the graduate level, admitting students with undergraduate degrees in mechanical or chemical engineering.

1.3 Defining Chemical Engineering

As with engineering, a clear and concise definition that captures the essence of chemical engineering is not readily available. The American Institute of Chemical Engineers (AIChE), the professional society of chemical engineers in the United States, defines chemical engineers as follows:

> Individuals who use science and mathematics, especially chemistry, biochemistry, applied mathematics, and engineering principles, to take laboratory or conceptual ideas and turn them into value added products in a cost effective, safe (including environmental) and cutting-edge processes.

This definition is clearly broad enough to cover all the activities that a chemical engineer engages in and yet not specific enough to distinguish it from other engineering disciplines. By specifically mentioning biochemistry, the definition also tends to show a bias toward biological or life sciences, which is not reflective of the occupation of a vast majority of chemical engineers.

An alternative, more concise definition of chemical engineering is presented by Morton M. Denn [12]:

> Chemical engineering is the field of applied science that employs chemical, physical and biochemical rate processes for the betterment of humanity.

The term *applied science* carries the meaning of taking laboratory or conceptual ideas to a larger scale. Chemical engineering differs from other engineering in having the scientific basis in chemistry *in addition* to physics and mathematical sciences. The concept of *rate processes* is at the heart of the previous definition. Chemical engineering, according to this definition, is the field dealing with the rates of physicochemical processes involving transformations of molecular species. The definition places emphasis on processes and the rates thereof. Although the process or the rate of transformation is of critical importance, affecting the economics of the engineering endeavor, it is ultimately the result of the transformation that is usually the desired outcome. In other words, the consumer or the society is interested in the product that can provide a needed service and not in the specifics of how that product is obtained.

The benefit or betterment of humanity is a common theme in most of the definitions of engineering. As previously mentioned, the term is too general, and engineering is not the only occupation working for the betterment of humanity. Despite the shortcomings, a combination of both definitions does convey the essence of chemical engineering as a profession. An individual can be identified as a chemical engineer, if the following descriptors can adequately describe his/her activities:

- The individual is engaged in an engineering enterprise; that is, he/she is working to apply scientific knowledge to make products, both tangible and intangible, commercially available to the general society.
- The engineering enterprise is based on the transformation of species, involving restructuring of bonds (forces) between them. This restructuring is typically at the atomic level, that is, involving chemical

> reactions yielding products that are distinct from initial species. However, the transformation may simply involve separations, or restructuring of physical bonds between different species. No new molecular species are generated in such transformations. In other words, the transformations involve altering the affinities between elemental and/or molecular species.

Chemical engineering thus deals generally with systems where chemical reactions take place. These chemical reactions are invariably accompanied by physical separations. These physical separations, as explained in subsequent chapters, often play the dominant role in determining the economics of the process. Chemical engineers use their knowledge of science and mathematics to ensure that laboratory reactions and separations are scaled up to the industrial level.

Historically, one can argue that individuals who fermented various brews were the earliest chemical engineers, predating even civil engineers. However, chemical engineering came into being as a distinct profession toward the end of the 19th century and beginning of the 20th century [12]. The development of the discipline resulted from increased demand for chemicals and fuels for both peacetime (fertilizers, consumer goods) and wartime (explosives) activities. Many of the technologies and products developed during the world wars led to additional industrial chemicals. Subsequent developments and societal needs have seen chemical engineering encompass myriad industries ranging from semiconductor, textile, pharmaceutical, agriculture, and food to energy, biotechnology, and medicine [13]. Chemical engineering is a highly versatile field full of challenges and opportunities in practically all facets of human activity.

1.4 Roles and Responsibilities of a Chemical Engineer

A chemical engineer performs a number of different tasks in the chemical industry or an organization engaged in chemical business. The tasks and responsibilities of the chemical engineer that are described in this section are based on the concept of engineering as the transformation process from science to technology.

Consider the case of manufacturing a new chemical product or a new manufacturing process for an existing chemical product. Going from conceptualization to commercial operation is an extremely complex and involved process, and any such undertaking typically moves through the following stages [14]:

1. Conceptualization and inception
2. Preliminary economic evaluation
3. Development of data needed for final design
4. Final economic evaluation
5. Detailed design
6. Procurement and construction
7. Start-up and trial runs
8. Production

A simplified flowchart for the various stages of a typical project are shown in Figure 1.7.

The first stage in this development consists of conceptualization of the idea for the new product or process, based on existing technical literature. Before an organization commits financial and other resources to any new venture, it conducts a preliminary economic analysis to gauge the economic viability of the venture. If the venture is judged to be economically attractive, the organization proceeds to the next stage: gathering the necessary information. Extensive laboratory experimentation is generally conducted to demonstrate the feasibility of the idea on the bench scale. Subsequent investigations are conducted over larger scales to ensure repeatable and reproducible results. An integrated process is developed for commercialization identifying the raw materials and processing steps, including those necessary for the treatment of effluents.

The key step, or possibly the entire process, is operated at pilot level to confirm the feasibility. The organization typically conducts a refined,

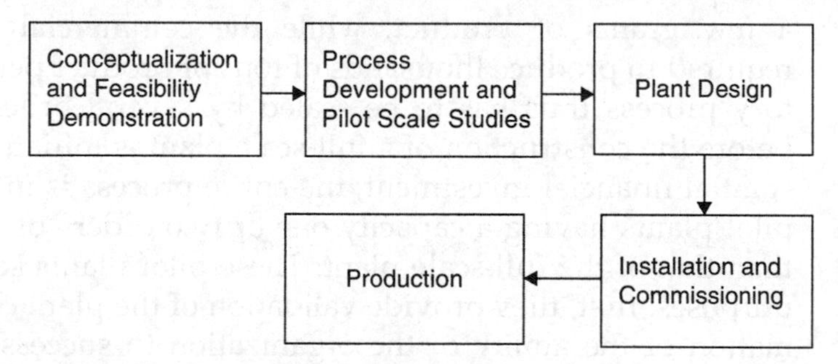

Figure 1.7 Engineering a chemical plant.

accurate economic analysis using the gathered data. If the project is still commercially attractive, a full-scale plant is designed using scientific and engineering principles and experimental data. The necessary equipment is fabricated and procured from vendors and installed, and the plant is commissioned. Modifications and fine-tuning of processes and operations is invariably required before the plant starts operating consistently, producing the desired product commercially.

A chemical engineer is involved in all of these activities. Depending on his/her responsibilities, the role of the engineer may include one or a combination of the following:

- *Research and development engineer:* An engineer who is involved in the initial stages of the diagram shown in Figure 1.7 is called a *research and development (R&D) engineer*. The research component of the position typically involves working with a chemist to investigate the key concept at the bench scale. The development component involves conceptualization of the process based on the key step. The engineer is responsible for identifying the sequence of upstream and downstream processing steps, evaluating alternatives, and conducting investigations on larger scales. Depending on the organization and the complexity of the process, an engineer may be engaged only in research, may conduct only developmental work, or may do both. The R&D engineer designs and executes experimental work, supervises staff, coordinates the data collection and analysis, and handles knowledge interchange among the members of the project team, including the chemist and engineering team. The R&D engineer is essentially the subject matter expert in the team, developing solutions to various challenges. A graduate degree (preferably Ph.D.) is highly desirable for an individual functioning as an R&D engineer.
- *Pilot plant engineer:* The laboratory investigations typically yield only a few grams of product, while the commercial process may be required to produce thousands of tons of product per day. The laboratory process thus has to be scaled by several orders of magnitude. Before the construction of a full-scale plant is initiated, requiring substantial financial investment, the entire process is inevitably tested in pilot plants having a capacity one or two orders of magnitude lower than that of the full-scale plant. These pilot plants serve two primary purposes: first, they provide validation of the plant design and confirmation of the ability of the organization to successfully operate the process on a large scale; second, they help identify the challenges that

arise with the scale-up of the process. For example, heating or cooling of material is generally not an issue on bench scale, but effective energy transfer can become a challenge on a larger scale. Similarly, whereas separation of solids by filtration can be a straightforward matter in the laboratory, it can limit the process efficiency and prove to be a bottleneck when scaled up. The pilot plant operation provides information about such potential technical problems and provides an opportunity to devise and test workable solutions to these problems. Pilot plant engineers run these plants, discovering potential problems, and devising, testing, and perfecting solutions to these problems before they manifest in the production plant.

- *Design engineer:* Sizing of the process equipment and specifying the operational conditions for the process are among the responsibilities of the design engineer. A design engineer, in the chemical engineering field, is mostly a *process design engineer*[4] who performs detailed calculations of energy and material flows, specifies equipment capacities, and determines the layout of the plant. The design engineer interfaces with the R&D and pilot plant engineers to obtain data and information and interfaces with vendors and fabricators to finalize process equipment specifications. He/she also typically develops cost estimates for the process. The tasks of fabrication of equipment, infrastructure development, and installations are typically executed by mechanical, civil, and electrical engineers. The design engineer interacts with these engineers and suppliers to ensure that the installed plant conforms to the design specifications.
- *Start-up, or commissioning, engineer:* The engineer involved in the initial start-up of a new plant is the start-up/commissioning engineer. He/she has the responsibility to ensure that various process units function as designed and the plant is able to deliver the product having the specified purity and quality at the rated design capacity. Once the initial problems are sorted out and the plant is operating smoothly, the commissioning engineer hands over the responsibility to the production operating personnel.
- *Manufacturing/production engineer:* A manufacturing/production engineer provides support to the manufacturing operation, monitoring daily operations to ensure that operations conform to the process design. He/she investigates process deviations, troubleshoots

4. In other fields, a design engineer may be a *product* design engineer, one who creates different product designs, such as a new phone or an app.

problems, and continually explores opportunities to increase profitability through capacity enhancement, efficiency improvements, and cost minimizations. The manufacturing/production engineer is responsible for plant safety, product quality and reliability, and costs for operations under his/her charge. Additionally, the manufacturing/production engineer works collaboratively with operations, engineering, maintenance, quality, and other departments to ensure plant reliability and production goals are met. Some plants may employ individuals other than chemical engineers to oversee the operation of the plant and its production. In this case, they may have a chemical engineer associated with the plant; such *plant engineers* are responsible for improving the productivity of the process, optimizing operations, and troubleshooting any problems that affect the operation of the plant. A facility may have a number of engineers who respond to such issues, and they may be called *technical services engineers*. All process plants also employ *maintenance engineers*, who ensure continuity of operation of various pieces of equipment through routine maintenance and repair of broken equipment. These engineers are typically mechanical or electrical engineers, and the production, plant, and technical services engineers interact closely with them.

Smaller organizations or simple processes may not require distinct personnel to perform the different jobs described. The same individual may perform multiple tasks, bearing the responsibility from laboratory investigations through commissioning. Conversely, larger organizations or complex processes may have separate and multiple individuals for each step. Commercialization of the process is always a team task, with the R&D engineer, pilot plant engineer, design engineer, and commissioning engineer exchanging information and refining/optimizing the process. An organization may also outsource many of the activities, contracting R&D with one entity, getting designs from specialist firms, and having plants commissioned by yet another entity.

Apart from these process- and plant-related jobs, many chemical engineers may work as sales or marketing engineers. Many companies need services of an individual to market their products, services, and equipment to other industries. This individual must have a thorough engineering understanding of the particular product/service being offered and is typically called a *sales engineer*, *technical sales engineer*, *marketing engineer*, or a variation thereof. A sales engineer must possess the necessary technical competence and must have excellent communication skills, a desire to interact with people, an ability to understand the client needs, and the ability to work closely with the team to develop products and services in response to these needs. Ability to cultivate relationships is an absolutely essential skill for such positions. A sales engineer invariably spends a considerable amount of time traveling, visiting clients in a territory typically assigned to him/her, and developing a customer base for the company.

Despite the different nature of responsibilities, all the engineering positions may be designated simply as *process engineers*, which is the most common title for the jobs advertised across the chemical industry. The next section describes briefly the various industries and other sectors of the economy that employ chemical engineers.

1.5 Employment of Chemical Engineers

The U.S. Department of Labor maintains an extensive database of labor and economic statistics through its Bureau of Labor Statistics (www.bls.gov). The career information about duties, education, training, pay, and current and future prospects for practically any occupation can be found in the *Occupational Outlook Handbook* (www.bls.gov/ooh) published by the Bureau. The handbook lists the myriad industries—chemicals, fuels, energy, food, drugs, electronics, clothing, life sciences, biotechnology, and many others—where chemical engineers find employment. It lists the median pay for the chemical engineering jobs to be $96,400 (for 2014) and projects a 2% annual growth in the number of jobs. The handbook very likely underestimates the actual number of chemical engineering jobs, with a more realistic number estimated by summing up the number of graduates over an average of 40 years [15]. A National Science Foundation (www.nsf.gov) estimate put the total number of working chemical engineers in 2002 across all occupations at closer to 200,000 [15].

Chemical process industries, composed of chemical and fuels manufacturing and associated R&D, engineering, and environmental services, are the biggest employers of chemical engineers [15]. The initial placement of new chemical engineering graduates conforms to this distribution as well. According to the 2015 initial placement survey of college graduates, conducted by the AIChE [16], nearly half of the graduates opted for industrial careers, and nearly a quarter of the bachelor's degree graduates chose to continue their education. Although a majority of those going to graduate study were continuing in the chemical engineering field, many chose different engineering fields or even medicine for further studies. For those opting for industrial careers, the chemicals sector was the dominant choice, followed by fuels, engineering services, and biotechnology, as shown in Figure 1.8.

This distribution was slightly different from the choices of 2014 graduates, of whom half still opted for industrial careers with chemicals as the dominant choice, but a significantly larger fraction opted for jobs in the fuels sector, and the biotechnology sector was not as dominant, lagging behind the food sector [17].

Traditionally, chemical engineers have pursued careers in the chemicals and fuels sectors of the industry. However, over the past two decades, increasingly greater numbers are being attracted by the pharmaceutical, biomedical, and microelectronics industries [18]. From a historical perspective, we can look back at chemical engineering since the beginning of the 20th century and recognize a few "golden ages" of the disciplines—times of unprecedented discoveries, developments, and industrial growth [19]. The 10 years from 1915 to 1925 and the decades of the 1950s and 1960s are time periods that exhibit such characteristics. It is, of course, impossible to predict the future; however, the current environment is characterized by possibly unlimited opportunities for chemical engineering due to developments in molecular biological sciences and computing and information technology. Chemical engineers, due to the breadth and depth of their preparation, are

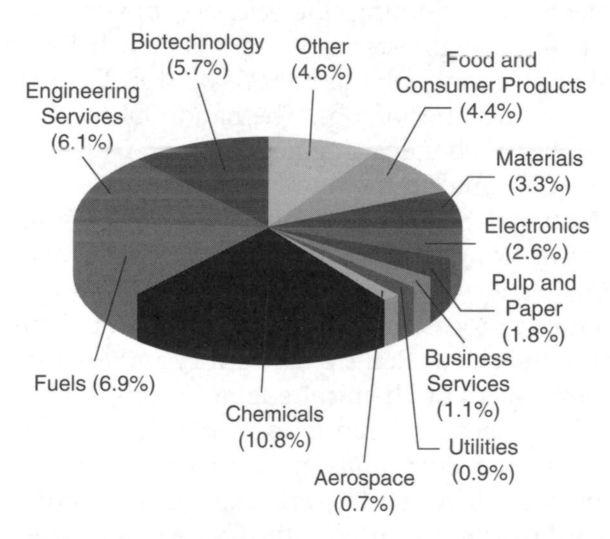

Figure 1.8 Initial placement of chemical engineering graduates in 2013–2014.

Source: CEP News Update, "AIChE's Initial Placement Survey: Where Is the Class of 2015?" *Chemical Engineering Progress*, Vol. 111, No. 12, 2015, pp. 5–6.

ideally positioned to enter this new golden age and find solutions to challenges in the areas of environment, resources, energy, food, health, and many other fields.

Salary and benefits data for chemical engineers are regularly collected by the professional societies of individuals associated with chemistry-related fields—the AIChE as well as the American Chemical Society (ACS, www.acs.org). Figure 1.9 shows the median initial starting salaries of inexperienced chemical engineers holding a bachelor's degree over the last 10 years [20]. These salaries have slightly outpaced inflation. As expected, the salaries also exhibit a slight dip coinciding with the overall downturn in the economy around 2009 to 2011.

The average starting salaries of chemical engineers are second only to those of petroleum engineers at the bachelor's level. However, chemical engineers command the highest salaries at the master's level [21]. The median salaries of chemical engineers having less than 6 years of experience, shown in Figure 1.10, are based on the data collected by AIChE through its biennial salary survey [22].

It can be seen that a graduate degree generally translates into higher salary for a chemical engineer, though this effect is not as pronounced for master's-level graduates as it is for those holding a doctorate. The differences in the salaries of bachelor's and master's degree holders were

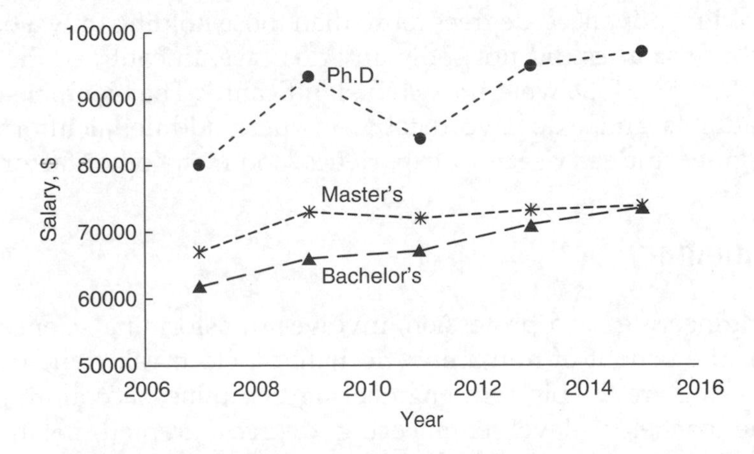

Figure 1.9 Median starting salaries of inexperienced chemical engineers with a bachelor's degree.

Source: Marchant, S., and C. Marchant, *Starting Salaries of Chemists and Chemical Engineers: 2014 Analysis of the American Chemical Society's Survey of New Graduates in Chemistry and Chemical Engineering*, American Chemical Society, Washington, D.C., 2015.

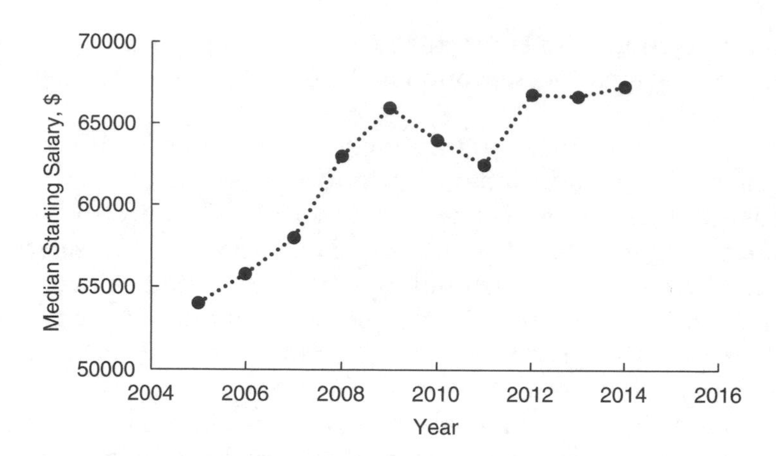

Figure 1.10 Median salaries of chemical engineers with less than 6 years of experience [22].

Source: *Chemical Engineering Progress*, "2007 AIChE Salary Survey," Vol. 103, No. 8, 2007, pp. 25–30; "2009 AIChE Salary Survey," Vol. 105, No. 8, 2009, pp. 26–32; "2011 AIChE Salary Survey," Vol. 107, No. 6, 2011, pp. S1–S13; "2013 AIChE Salary Survey," Vol. 109, No. 6, 2013, pp. S1–S17; "2015 AIChE Salary Survey," Vol. 111, No. 6, 2015, pp. S1–S20.

insignificant in 2015. Further, the economic downturn clearly impacted those holding advanced degrees more than those holding only a bachelor's degree, whose salaries did not show any decrease. In contrast, the salaries of those holding a Ph.D. were impacted significantly. The biennial surveys provide a valuable and extensive database, where additional information based on regions, industry sectors, experience, and many other factors is available.

1.6 Summary

Engineering, as a profession, involves transforming science into technology for the benefit of humanity. Mechanical, electrical, civil, and chemical engineering are the big four engineering disciplines, accounting for over half of the bachelor's-level engineering degrees granted in the United States. Chemical engineering is the branch of engineering dealing with the transformation of species at the molecular level. A chemical engineer is a versatile individual who can serve in different capacities in an organization, with responsibilities ranging from research to technical sales. Over half of the chemical engineering graduates enter the workforce in the industrial sector, with the chemical sector offering predominant employment opportunities.

Recent developments indicate a continual growth in opportunities for engineers in biological science–related fields. A chemical engineer, due to the breadth and depth of education and training, is ideally positioned to take advantage of these opportunities to play a major role in addressing challenges facing the society.

References

1. Besterfield-Sacre, M., C. J. Altman, and L. J. Shuman, "Characteristics of Freshman Engineering Students: Models for Determining Student Attrition in Engineering," *Journal of Engineering Education*, John Wiley and Sons, Vol. 86, No. 2, 1997, pp. 139–149.
2. Kemper, J. D., and B. R. Sanders, *Engineers and Their Profession*, Fifth Edition, Oxford University Press, New York, 2001.
3. Burghardt, M. D., *Introduction to the Engineering Profession*, Second Edition, Prentice Hall, Upper Saddle River, New Jersey, 1997.
4. Landis, R. B., *Studying Engineering: A Road Map to a Rewarding Career*, Fourth Edition, Discovery Press, Los Angeles, 2013.
5. Shackelford, J. F., *Introduction to Materials Science for Engineers*, Seventh Edition, Prentice Hall, Upper Saddle River, New Jersey, 2009.
6. Smith, H. S., *The World's Great Bridges*, Harper and Row, New York, 1965.
7. Hauck, G. F. W., "The Structural Design of the Pont du Gard," *Journal of Structural Engineering*, Vol. 112, No. 1, 1986, pp. 105–120.
8. Kenoyer, J. M., *Ancient Cities of the Indus Valley Civilization*, Oxford University Press, Oxford, U.K., 1998.
9. Chondros, T. G., "Archimedes life works and machines," *Mechanism and Machine Theory*, Vol. 45, No. 11, 2010, pp. 1766–1775.
10. Yoder, B. L., "Engineering by the Numbers," https://www.asee.org/papers-and-publications/publications/college-profiles/15EngineeringbytheNumbersPart1.pdf.
11. Muslih, I., D. B. Meredith, and A. S. Kuzmar, "Overlap Between Mechanical and Civil Engineering Undergraduate Education," American Society of Engineering Education, 2005 IL/IN Sectional Meeting, April 2005, Dekalb, Illinois, Paper D-T1–2.
12. Denn, M. M., *Chemical Engineering: An Introduction*, Cambridge University Press, Cambridge, U.K., 2012.
13. Solen, K. A., and J. N. Harb, *Introduction to Chemical Engineering: Tools for Today and Tomorrow*, Fifth Edition, John Wiley and Sons, New York, 2011.
14. Peters, M., K. Timmerhaus, and R. West, *Plant Design and Economics for Chemical Engineers*, Fifth Edition, McGraw-Hill, New York, 2012.

15. Self, F., and E. Eckholm, "Employment of Chemical Engineers," *Chemical Engineering Progress*, Vol. 99, No. 1, 2002, pp. 22S–25S.
16. CEP News Update, "AIChE's Initial Placement Survey: Where Is the Class of 2015?" *Chemical Engineering Progress*, Vol. 111, No. 12, 2015, pp. 5–6.
17. CEP News Update, "AIChE's Initial Placement Survey: Where Is the Class of 2014?" *Chemical Engineering Progress*, Vol. 110, No. 11, 2014, pp. 5–6.
18. Himmelblau, D. M., and J. B. Riggs, *Basic Principles and Calculations in Chemical Engineering*, Eighth Edition, Prentice Hall, Upper Saddle River, New Jersey, 2012.
19. Westmoreland, P. R., "Opportunities and Challenges for a Golden Age of Chemical Engineering," *Frontiers of Chemical Engineering Science*, Vol. 8, No. 1, 2014, pp. 1–7.
20. Marchant, S., and C. Marchant, *Starting Salaries of Chemists and Chemical Engineers: 2014 Analysis of the American Chemical Society's Survey of New Graduates in Chemistry and Chemical Engineering*, American Chemical Society, Washington, D.C., 2015.
21. National Association of Colleges and Employers, "Top-Paid Engineering Majors at the Bachelor's and Master's Levels," 2015, http://naceweb.org/s03182015/top-paid-engineering-majors.aspx.
22. *Chemical Engineering Progress*, "2007 AIChE Salary Survey," Vol. 103, No. 8, 2007, pp. 25–30; "2009 AIChE Salary Survey," Vol. 105, No. 8, 2009, pp. 26–32; "2011 AIChE Salary Survey," Vol. 107, No. 6, 2011, pp. S1–S13; "2013 AIChE Salary Survey," Vol. 109, No. 6, 2013, pp. S1-S17; "2015 AIChE Salary Survey," Vol. 111, No. 6, 2015, pp. S1–S20.

Problems

1.1 Discuss the similarities and differences between engineering and technology.

1.2 Search the literature and list 10 different definitions of engineering. Use any source, including those (websites, books) mentioned in the "References" section. Discuss comparative features of three alternative definitions. Which one do you prefer? Why?

1.3 How many engineers have graduated with a bachelor's degree over the last 5 years in the United States? Which disciplines exhibit the highest and lowest rates of growth? How do these compare to trends for chemical engineering?

1.4 Use the data available from the Bureau of Labor Statistics to compare projected growth in jobs for the four major engineering disciplines.

1.5 What are the trends in the placement of recent chemical engineering graduates over the past 5 years? Which sectors show higher growth rates? Which sectors show negative growth rates?

1.6 Which of the responsibilities of a chemical engineering position is more appealing to you? Why?

1.7 What were the two golden ages of chemical engineering according to reference 19?

1.8 Which of the emerging new applications of chemical engineering are attractive to you? What features make the opportunities in these fields desirable for you?

CHAPTER 2
Chemical and Allied Industries

The chemical industry is of strategic importance to the sustainable development of national economies.
—International Labor Organization[1]

Chemical engineers have traditionally found employment in the chemical and allied industries, and these industries continue to be the major employers of chemical engineers. The chemical and allied industries comprise one of the most important manufacturing sectors of a nation's economy. However, despite their significance, these industries are not well understood by the general public, partly because only a small fraction of the output of these industries is consumer product; the bulk of it is raw material for other industries. This chapter presents an overview of the chemical and allied industries with the objective of providing chemical engineering students an understanding of their most probable source of employment opportunities.

Section 2.1 describes the classification of the industries with a brief introduction to the systems used by the United States and other governments to monitor and analyze the economy. The chemical and related industries are described in sections 2.2 and 2.3, respectively, followed by a discussion of top chemical companies in section 2.4. Section 2.5 describes some of the important chemical products, and section 2.6 describes the general characteristics of the chemical industry. Readers will become better acquainted with the significance of the chemical and allied industries in a nation's economy, as well as have a greater appreciation of the indispensable role of chemicals in modern society.

2.1 Classification of Industries

Before venturing into the nature of chemical and allied industries and understanding its role, it is instructive to look at the classification system used by governments to analyze and track the various sectors of the nation's

1. International Labor Organization (www.ilo.org/global/industries-and-sectors/chemical-industries/lang--en/index.htm)

economy. The U.S. government developed a standardized system in the 1930s called the Standard Industrial Classification (SIC) system for classifying industries. The SIC system has also been adopted by other countries. Major business and industry sectors are represented by a four-digit numerical SIC code on the basis of common characteristics. These SIC codes are hierarchical, with the first two digits representing the major business/industry sector and the third and fourth digits representing subclassifications and specializations within the major sector (https://www.osha.gov/pls/imis/sic_manual.html).

The United States has been transitioning to a newer classification system—the North American Industry Classification System (NAICS)—since the late 1990s (www.census.gov/eos/www/naics). NAICS codes are six-digit numbers based on a top-down hierarchical structure similar to the SIC system. Each SIC code has a corresponding unique NAICS code in the newer system. Both the SIC and NAICS codes are used by various entities in the United States. The SIC and NAICS codes of industries of primary relevance to chemical engineers are shown in Table 2.1.

The manufacturing industries have been assigned codes beginning with 31, 32, or 33 under the NAICS. Chemical and allied products belong to the manufacturing industries sector, with chemical products carrying the NAICS code beginning with 325. The combined contributions of chemical, petroleum/coal, paper, and rubber products amounted to nearly $670 billion in the United States, equal to nearly 33% of all manufacturing in 2012. Chemical products alone (SIC code 28/NAICS code 325) accounted for almost 18% of the manufacturing sector.[2]

Chemical Products are further subclassified into Basic Inorganic Chemicals, Industrial Gases, Basic Organic Chemicals, Fertilizer Products, Polymer Products, Pharmaceuticals, and many others. Some of the

Table 2.1 Classification of Chemical and Related Industries

SIC System		NAICS	
Code	**Industry**	**Code**	**Industry**
26	Paper and Allied Products	322	Paper Products
28	Chemical and Allied Products	324	Petroleum and Coal Products
29	Petroleum and Coal Products	325	Chemical Products
30	Rubber and Misc. Plastics Products	326	Plastics and Rubber Products

2. U.S. Department of Commerce, Bureau of Economic Analysis Data.

important outputs of the Chemical Industry (NAICS Code 325) and Related Industries (NAICS Codes 322, 324 and 326) are briefly described in the following sections.

2.2 The Chemical Industry

The chemical industry manufactures a vast number of chemicals to serve the needs of the society. Chemicals sharing common characteristics are grouped together and classified under the same NAICS codes for analyzing their economic impact. The chemical products are broadly classified into seven categories: Basic Inorganic Chemicals, Industrial Gases, Basic Organic Chemicals and Petrochemicals, Fertilizer Products, Polymer Products, Pharmaceutical Products, and Other Chemical Products. A brief description of each of these classes of chemical products is presented in the following subsections.

2.2.1 Basic Inorganic Chemicals

Chemicals produced in bulk quantities and used primarily in subsequent industrial processes are termed as *basic chemicals*. Basic inorganic chemicals are, as the name indicates, non-carbon-based compounds, though carbon dioxide and inorganic carbonates are included under this category. Nearly 40% of the top 50 chemicals are basic inorganic chemicals [1]. Typically, seven of the top 10 chemicals produced worldwide are inorganic: sulfuric acid, nitrogen, oxygen, chlorine, phosphoric acid, ammonia, and sodium hydroxide. Nitrogen and oxygen are categorized under Industrial Gases, whereas ammonia is categorized under Fertilizer Products. The other four chemicals are considered Basic Inorganic Chemicals. Almost all of the inorganic chemicals are industrial products; that is, they are used in production of other chemicals and consumer products. For example, the largest use of sulfuric acid is in phosphate-based fertilizer production. Similarly, caustic soda (sodium hydroxide) and chlorine are used in the manufacture of organic chemicals and the pulp and paper industry.

2.2.2 Industrial Gases

Except for chlorine, which is considered a basic inorganic chemical, other industrially significant gases have their own distinct category. The most important of industrial gases are oxygen, nitrogen, hydrogen, and carbon dioxide. Oxygen and nitrogen are obtained from air primarily through cryogenic liquefaction and distillation. The use of nitrogen (and of hydrogen) is dominated by ammonia manufacture. Nitrogen is also used in enhanced oil recovery (EOR) and maintenance of inert atmosphere in processes, and the other significant application of hydrogen is in adjusting the carbon-to-hydrogen ratio in hydrocarbons, particularly transportation fuels. Pure oxygen is used in chemical manufacturing, including that of metals, and in medical applications. Carbon dioxide, as an industrial gas, is formed as a by-product of hydrogen manufacture. Its dominant use is in refrigeration, the food industry, and manufacturing chemicals. Figure 2.1 provides a very broad overview of the basic inorganic chemical and industrial gas products.

2.2.3 Basic Organic Chemicals and Petrochemicals

Almost all of the organic chemicals produced by the chemical industry are obtained from seven basic chemicals: methane, ethylene, propylene, butadiene, benzene, toluene, and xylene. Methane is the primary constituent of natural gas. Although some minor sources are available for the rest of the basic organic chemicals, they are invariably derived from petroleum, and chemicals derived from petroleum are collectively termed *petrochemicals*. Ethylene is the largest-volume organic chemical produced worldwide, followed by propylene. The largest use of these chemicals is in the manufacture of polymers, such as polyethylene, polypropylene, polybutadiene and rubbers, and polyesters (polyethylene terephthalate [PET] and polybutylene terephthalate

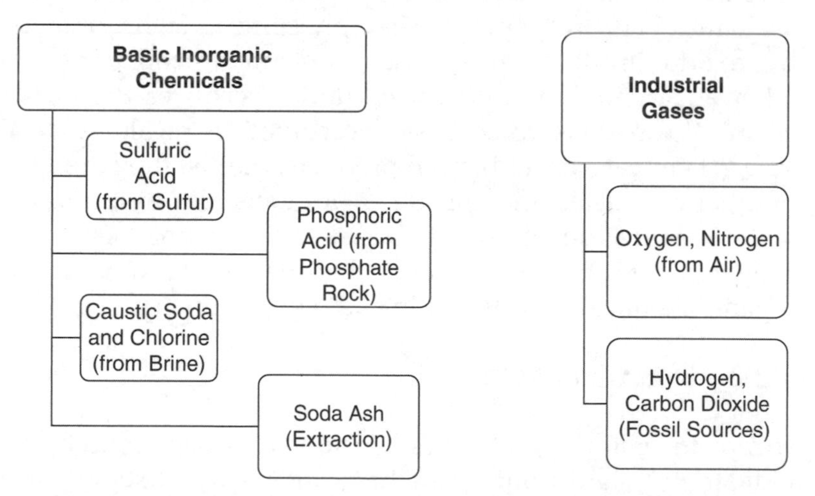

Figure 2.1 Major basic inorganic chemicals and industrial gases.

[PBT]). The other important chemicals produced from these chemicals include ethylene oxide and vinyl chloride, propylene oxide and isopropyl alcohol, cyclohexane, toluene diisocyanate, and phthalic anhydride. Many of these derived chemicals are, in turn, used in manufacture of other organic compounds, including polymers such as polyvinyl chloride (PVC) and polyurethanes.

Figure 2.2 provides an overview of the basic organic chemicals and petrochemical products. The top level in the hierarchical representation shows the different sources of the basic chemicals. As mentioned previously, petroleum (crude oil) is the primary source for the chemicals; however, these chemicals, particularly methane, are also obtained from coal. The middle level in the figure shows the seven basic organic chemicals; methane is separate primarily to indicate that it is predominantly obtained from natural gas. The lower level provides a brief glimpse into the world of petrochemicals: the intermediates such as butadiene, ethylene dichloride, acrylonitrile, formaldehyde, and countless others that are precursors to a vast number of chemical products that go into objects used universally by practically every individual.

2.2.4 Fertilizer Products

Fertilizer industries are vital components of a nation's economy; however, the significance of fertilizer products cannot be measured merely by the

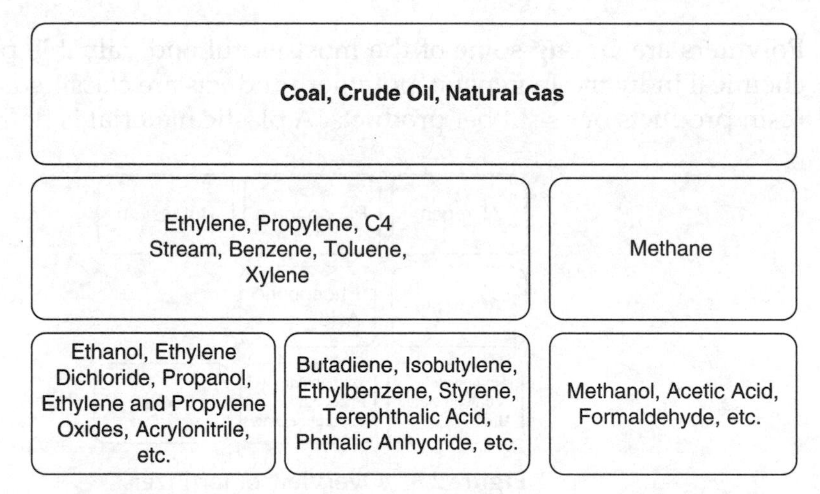

Figure 2.2 Overview of basic organic chemicals and petrochemical products.

economic value they add. The sustainability of food production is critically dependent on the availability of fertilizers. In that sense, fertilizer production is the most important endeavor of the chemical industry. Fertilizer products utilize the basic inorganic and organic chemicals and industrial gases as described earlier. Fertilizer products belong primarily to one of the two major classes: nitrogenous fertilizers and phosphatic fertilizers. Nitrogenous fertilizers include products such as ammonium nitrate, ammonium sulfate, and urea; phosphatic fertilizers include ammonium phosphates and superphosphates. Figure 2.3 provides an overview of the fertilizer products that are indispensable for satisfying the food demands of the ever-increasing human population.

Phosphatic fertilizers are based on phosphoric acid formed by reacting the raw material phosphate rock with sulfuric acid, a basic inorganic chemical, as previously mentioned. Nitrogenous fertilizers are based on ammonia, which itself is formed from the industrial gases nitrogen and hydrogen. Most fertilizer products available in the market are characterized by a three-number label, with the numbers representing the elemental content with respect to nitrogen, phosphorus, and potassium (N-P-K).[3] The desired level of potassium in the particular product is obtained by blending appropriate quantity of potash (generally potassium chloride, though the term *potash* is variously used to refer to potassium carbonate, hydroxide, chloride, or oxide [1]) into the product.

2.2.5 Polymer Products

Polymers are among some of the most useful and valuable products of the chemical industry. In general, polymer products are classified as plastics and resin products or as rubber products. A plastic material is defined as a solid,

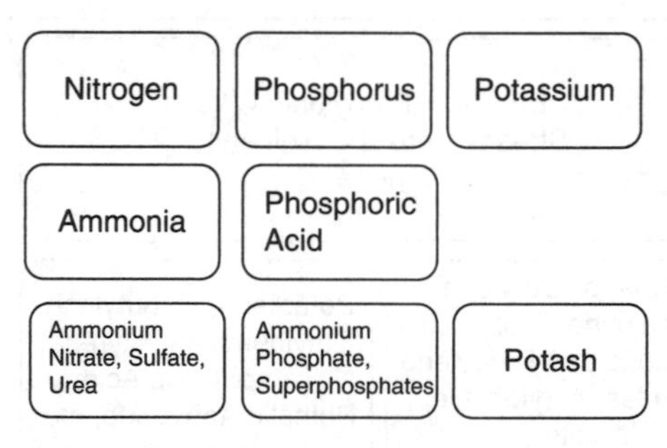

Figure 2.3 Overview of fertilizer products.

3. The numbers for P and K actually refer to the percentages of P_2O_5 and K_2O, respectively.

high-molecular-weight, polymerized organic substance and typically involves liquid-phase manufacturing or processing of the material [2]. Plastic and resin products include chemicals such as polyolefins (polyethylene/polypropylene), polyesters (PET, PBT), polyamides (nylons), and formaldehyde-based resins. Rubber products are, similarly, polymers having a high molecular weight, but in contrast to plastic and resin products, they exhibit elastic behavior and often are called *elastomers* [1]. Rubber products include synthetic rubber, latex, nitrile, silicone, and other rubbers.

The polymer product sector is inevitably linked to the basic organic chemicals and the petrochemicals. Most of the polymers are formed either through addition reactions (involving addition of the monomer molecule to the polymer chain) or condensation reactions (involving reaction groups belonging to two different monomer units). Most of the polyolefins, such as polyethylene, polypropylene, and PVC, are formed through the addition polymerization reactions. Such polymers typically soften and melt without decomposition upon heating and can be resolidified into new forms without losing their polymer characteristics. These polymers are called *thermoplastics* [1]. Phenolic resins, such as phenol-formaldehyde resins, and polyesters, such as PET, are formed by condensation reactions. These polymers are characterized by strong chemical bond crosslinking of units. They typically decompose upon heating and, unlike thermoplastics, cannot be reconstituted upon cooling. Figure 2.4 shows some of the

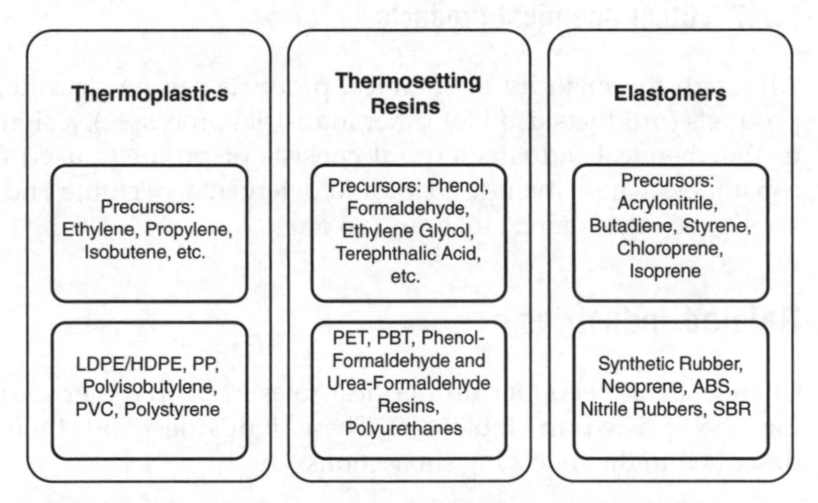

Figure 2.4 Classification and examples of polymer products. ABS, acrylonitrile-butadiene-styrene; LDPE, low-density polyethylene; HDPE, high-density polyethylene; PBT, polybutylene terephthalate; PET, polyethylene terephthalate; PVC, polyvinyl chloride; SBR, styrene butadiene rubber.

important polymer products. The figure offers only a brief glimpse into the world of polymers and does not include several other products of equal significance.

The polymer sector of the chemical industry includes the manufacture not only of the polymer (polyester, for example) but also of fibers and filaments based on these polymers. Polymers are highly versatile chemicals with remarkable properties that allow them to be shaped into thin films, flexible nets, rigid furniture, cloth, and countless other products that make them so ubiquitous in modern society.

2.2.6 Pharmaceutical Products

The pharmaceutical sector of the chemical industry involves the manufacture of (1) uncompounded medicinal compounds and (2) pharmaceutical formulations (tablets, ointments, etc.) that can be administered as doses. As with the fertilizer product sector, the significance of this sector goes beyond the mere monetary value to the economy. Pharmaceutical products are generally produced in much smaller quantities than other products previously described. However, these are value-added products that command a substantially higher price. The products also have a much more complex structure than the simple chemicals discussed earlier.

2.2.7 Other Chemical Products

Although the majority of chemical products can be classified as industrial products (products used for other industrial processes), a significant fraction of the chemical industry output consists of products used directly by the consumers. These include soaps and detergents, perfume and cosmetics, flavors, pesticides, paints, inks, and so on.

2.3 Related Industries

Chemical engineers perform critical roles in many other industries, including those listed in Table 2.1. These industries and their products are described in the following subsections.

2.3.1 Paper Products

Industries classified under NAICS code 322 are involved in the manufacture of paper products. These industries include pulp mills that convert wood and biomass to pulp and paper mills that consume this pulp to manufacture various types of paper, newsprint, paperboard, and consumer products such as cardboard boxes, paper bags, sanitary paper, and many others. These industries use large quantities of chemical products and involve processes similar to those of the chemical industries.

2.3.2 Petroleum and Coal Products

Petroleum, the result of transformations of organic matter over millions of years, is a mixture of a large number of compounds. This crude material has to be refined and separated into valuable products such as transportation fuels, basic organics, and other chemicals. Petroleum-derived products listed under NAICS code 324 include gasoline, jet fuel, kerosene, fuel oils, lubricating oils and greases, and asphalt and asphalt products. Coal products include coke oven and blast furnace products (except for those made in steel mills) such as coke, crude oil, and so on. The growth of chemical engineering as a distinct engineering discipline is in large part due to these needs in petroleum processing and refining [2]. Processing of conventional fossil resources such as petroleum and coal, as well as nonconventional resources such as tar sands and oil shale, creates many employment opportunities for chemical engineers.

2.3.3 Plastics and Rubber Products

Chemical engineers find employment not only in the industries that manufacture polymers but also in those that convert these polymers into finished products ranging from grocery bags, plastic bottles, and films to tires, hoses and tubing, and rubber belts. These industries are classified under NAICS code 326, unlike the polymer products described in section 2.2.5 that are classified under code 325.

2.3.4 Other Related Industries

Apart from these three most closely related industries, chemical engineering principles and operations are also encountered in many other industries and processes, providing employment opportunities to chemical engineers. These industries include ceramic and glass industries, semiconductor industries, food products and processing, energy and nuclear industries, environmental and pollution control undertakings, and many others.

2.4 Top 50 Chemical Companies

Table 2.2 shows the global top 50 chemical companies based on the most recent *C&E News* data [3]. The top 50 U.S. chemical companies are shown in Table 2.3 [4].

The following important observations can be made regarding the chemical industry based on the information from these two tables, similar data from prior years, as well as from the websites of the companies:

- *The chemical industry is a global enterprise:* The global top 50 list may be dominated by the companies headquartered in the United States, Germany, and Japan, but all the continents and a large number of countries have representatives in the list. These include developed

Table 2.2 Global Top 50 Chemical Companies*

Rank					
2015	2014	Company	2015 Sales $ Billions	Chemical Sales as a % of Total Sales	Headquarters Country
1	1	BASF	63.7	81.5	Germany
2	2	Dow Chemical[a]	48.8	100	United States
3	3	Sinopec	43.8	13.9	China
4	4	SABIC	34.3	87.0	Saudi Arabia
5	6	Formosa Plastics	29.2	63.9	Taiwan
6	9	Ineos	28.5	100	Switzerland
7	5	ExxonMobil	28.1	10.8	United States
8	7	LyondellBasell Industries	26.7	81.5	Netherlands
9	11	Mitsubishi Chemical	24.3	77.1	Japan
10	8	DuPont[a]	20.7	82.4	United States
11	13	LG Chem	18.2	100	South Korea
12	15	Air Liquide	17.3	95.3	France
13	17	Linde	16.8	84.5	Germany
14	16	AkzoNobel	16.5	100	Netherlands
15	21	Toray Industries	15.5	89.3	Japan
16	20	Evonik Industries	15.0	100	Germany
17	24	PPG Industries	14.2	92.9	United States
18	14	Braskem	14.2	100	Brazil
19	23	Yara	13.9	100	Norway
20	—	Covestro	13.4	100	Germany

Rank		Company	2015 Sales $ Billions	Chemical Sales as a % of Total Sales	Headquarters Country
2015	2014				
21	18	Sumitomo Chemical	13.3	76.6	Japan
22	22	Reliance Industries	12.9	29.8	India
23	25	Solvay	12.3	100	Belgium
24	10	Bayer	11.5	30.2	Germany
25	19	Mitsui Chemicals	11.1	100	Japan
26	29	Praxair	10.8	100	United States
27	31	Shin-Etsu Chemical	10.6	100	Japan
28	26	Lotte Chemicals	10.4	100	South Korea
29	32	Huntsman Corp.	10.3	100	United States
30	33	Syngenta	9.9	74.0	Switzerland
31	28	DSM	9.9	100	Netherlands
32	38	Air Products & Chemicals	9.9	100	United States
33	39	Eastman Chemical	9.6	100	United States
34	27	Chevron Phillips Chemical	9.2	100	United States
35	41	Mosaic	8.9	100	United States
36	35	Lanxess	8.8	100	Germany
37	34	Borealis	8.5	100	Austria
38	43	Arkema	8.5	100	France
39	36	Asahi Kasei	8.4	51.7	Japan
40	37	Sasol	8.3	57.1	South Africa
41	30	SK Innovation	8.2	19.2	South Korea
42	42	DIC	7.1	100	Japan
43	45	Hanwha Chemical	7.1	100	South Korea
44	—	Lubrizol	7.0	100	United States
45	49	Ecolab	6.9	50.7	United States
46	47	Indorama	6.9	100	Thailand
47	50	Johnson Matthey	6.5	39.8	United Kingdom
48	—	Honeywell	6.5	16.8	United States
49	40	PTI Global Chemical	6.4	54.6	Thailand
50	—	PotashCorp	6.3	100	Canada

*Based on data from *C&E News* [3].

[a]Dow Chemical Company and E.I. du Pont de Nemours announced plans for merger in December 2015. When completed, it will create DowduPont, one of the largest chemical companies ever.

Table 2.3 Top 50 U.S. Chemical Companies*

Rank				
2015	2014	Company	2015 Sales $ Billions	Headquarters State
1	1	Dow Chemical	48.8	Michigan
2	2	ExxonMobil	28.1	Texas
3	3	DuPont	20.7	Delaware
4	4	PPG Industries	14.2	Pennsylvania
5	6	Praxair	10.7	Connecticut
6	7	Huntsman	10.3	Texas
7	8	Air Products	9.9	Pennsylvania
8	9	Eastman Chemical	9.6	Tennessee
9	5	Chevron Phillips	9.2	Texas
10	10	Mosaic	8.9	Minnesota
11	13	Lubrizol	7.0	Ohio
12	11	Ecolab	6.8	Minnesota
13	12	Honeywell	6.5	New Jersey
14	—	Chemours	5.7	Delaware
15	14	Celanese	5.7	Texas
16	15	Dow Corning	5.6	Michigan
17	18	Monsanto	4.8	Missouri
18	21	Westlake Chemical	4.5	Texas
19	20	CF Industries	4.3	Illinois
20	16	Hexion	4.1	Ohio
21	17	Trineso	4.0	Pennsylvania
22	19	Occidental Petroleum	3.9	Texas
23	28	Albemarle	3.7	Louisiana
24	22	Ashland	3.4	Kentucky
25	23	FMC Corp.	3.3	Pennsylvania
26	26	W.R. Grace	3.1	Maryland
27	25	Cabot Corp.	2.9	Massachusetts
28	24	Axiall	2.7	Georgia
29	27	Momentive	2.3	New York
30	37	Olin	2.1	Missouri
31	29	Newmarket	2.1	Virginia
32	31	H.B. Fuller	2.1	Minnesota

Rank				
2015	2014	Company	2015 Sales $ Billions	Headquarters State
33	34	Stepan	1.8	Illinois
34	30	Chemtura	1.7	Pennsylvania
35	33	Americas Styrenics	1.6	Texas
36	32	Cytec Industries	1.5	New Jersey
37	36	Kronos Worldwide	1.3	Texas
38	35	Sigma-Aldrich	1.2	Missouri
39	39	Ferro Corp.	1.1	Ohio
40	38	Kraton Polymers	1.0	Texas
41	44	Innospec	1.0	Colorado
42	45	Koppers	0.97	Pennsylvania
43	43	Ingevity	0.97	South Carolina
44	40	Reichhold	0.95	North Carolina
45	—	Arizona Chemical	0.80	Florida
46	47	PolyOne	0.80	Ohio
47	46	Innophos	0.79	New Jersey
48	48	Emerald Performance Materials	0.70	Ohio
49	50	Mineral Technologies	0.62	New York
50	49	Omnova	0.61	Ohio

*Based on data from *C&E News* [4].

nations such as France, Switzerland, and Netherlands and emerging economies such as China, Brazil, and India. Further, while the companies may have headquarters in a country, they are almost certain to have a significant global presence. For example, companies such as BASF and SABIC have substantial operations in the United States, and conversely, U.S. companies such as Dow and DuPont have production and research and development facilities in many countries. Geographically for the United States, the largest of the chemical companies are located near the east coast and south (states bordering the Gulf of Mexico), with a substantial presence in the Midwest region (Minnesota, Missouri, Michigan) as well. These regions also boast of the largest concentration of the oil and gas industry, and the connection between the two industries should be clear. After all, the chemical industry obtains most of its feedstock for organic chemicals from the oil and gas industry.

- *The chemical industry is mature:* Practically all companies in the 2015 global top 50 list also appear in the 2014 list and in the rankings for prior years (not shown in the table). Further, for the most part there are no dramatic changes in the rankings of the companies. The few new entrants that appear are generally near the bottom of the rankings. Similar observation can be made regarding the list of top 50 U.S. chemical companies.
- *Chemical operations constitute a small fraction for oil companies:* Most large oil companies—ExxonMobil, Sinopec, BP, Shell, and so on—are also large chemical manufacturers. However, the chemical business for these companies is dwarfed by the revenues from oil and gas operations. As seen from Table 2.2, chemical business contributes typically less than 10% of the revenue for these companies.
- *The chemical companies are well diversified:* The chemical companies have a wide-ranging portfolio of products. Companies such as DuPont and Dow produce chemicals that have general applicability across many industrial sectors as well as specific products for applications in particular industries ranging from automotive to animal feed, food and beverage to hospital care, and many others.

2.5 Important Chemical Products

The United States occupies the top position among chemical manufacturers, accounting for nearly one-fifth of the chemicals produced worldwide [5]. The total value of the chemical goods produced in the United States in 2010 was $701 billion, and the chemical manufacturing sector contributed the largest fraction to the U.S. exports. Table 2.4 shows the 2010 production volumes of some of the important chemicals [6].

The following subsections provide an overview of the manufacturing technologies for some of these important chemicals and their applications.

2.5.1 Sulfuric Acid

Sulfuric acid is by far the largest-volume chemical produced by the chemical industry, and sulfuric acid production is an excellent indicator of a nation's economy. The U.S. annual production of sulfuric acid is ~36 million tons [5]. The *contact process* for the manufacture of sulfuric acid involves catalytic oxidation of sulfur dioxide and is a classic example of a chemical process [2]. Figure 2.5 shows a simplified process flow diagram

Table 2.4 2010 Production Volumes of Important Chemicals*

Product Category	Chemical	Production Volume, Million Metric Tons
Basic Inorganic Chemicals	Sulfuric acid	32.6
	Phosphoric acid	9.4
	Chlorine	9.7
	Caustic soda (sodium hydroxide)	7.5
	Soda ash (sodium carbonate)	10.6
Industrial Gases	Nitrogen	31.6
	Oxygen	26.5
	Carbon dioxide	7.9
	Hydrogen	3[a]
Ammonia and Fertilizers	Ammonia	10.3
	Ammonium phosphates	12.1
	Ammonium nitrate	6.9
	Urea	5.1
Basic Organic Chemicals	Ethylene	24.0
	Propylene	7.8
	C4 hydrocarbons (butylenes and butadiene)	6.5
	Benzene	6.0
	Xylenes	9.7

*Based on data from the U.S. Department of Energy [6].

[a]Does not reflect captive production for manufacturing other chemicals.

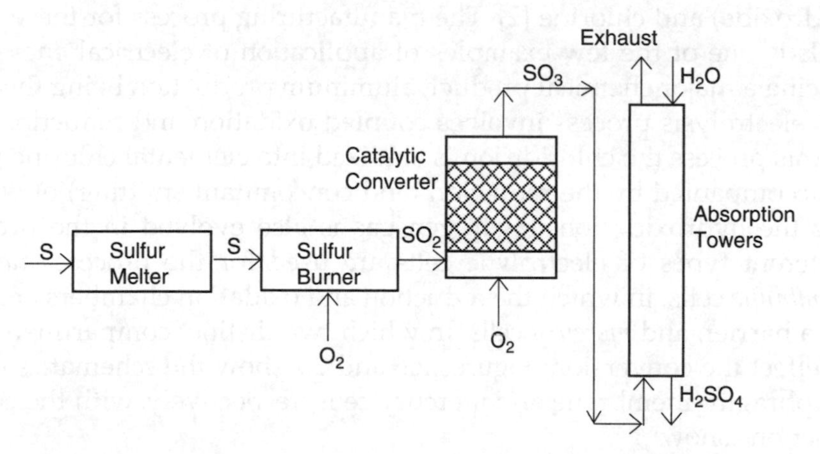

Figure 2.5 Sulfuric acid manufacture process flowsheet.

for the contact process. Molten sulfur is fed to the sulfur burner where it is converted to sulfur dioxide by burning it with air. The key step in the process is the conversion of the sulfur dioxide to sulfur trioxide, a reaction that can be carried out economically by using a cheap vanadium pentoxide (V_2O_5) catalyst.

The design of the catalytic reactor for conducting this reaction represents a classical application of chemical engineering principles. Absorption of sulfur trioxide in water yields sulfuric acid; however, the absorption towers are operated such that the sulfur trioxide exiting the catalytic reactor is first contacted with highly concentrated (>98%) sulfuric acid rather than water. Direct absorption of sulfur trioxide in water is ineffectual, as contacting sulfur trioxide with water results in the formation of acid mist [2]. Typically, two absorption towers are used in most processes; this is known as the *double contact double absorption* (DCDA) process.

Sulfur, raw material for the process, used to be obtained from sulfur mines and sulfur-bearing porous limestone. At the present time, sulfur associated with fossil resources (sour crude oil, high sulfur coal) as well as pyrite ores supply the sulfur demand for sulfuric acid. A very large percentage of sulfuric acid produced is used in the manufacture of phosphate fertilizers, with other uses consisting of manufacture of a variety of chemicals, petroleum purification, and metal pickling [1].

2.5.2 Caustic Soda and Chlorine

Electrolysis of brine (sodium chloride solution) yields two inorganic chemicals that are major outputs of the chloralkali industries: caustic soda (sodium hydroxide) and chlorine [2]. The manufacturing process for these two chemicals is one of the few examples of application of electrical energy for producing a major chemical product, aluminum production being the other one. An electrolysis process involves coupled oxidation and reduction reactions; in this process the chloride ion is oxidized into elemental chlorine gas, which is accompanied by the reduction (and concomitant splitting) of water yielding the hydroxide ion. Hydrogen gas is also evolved in the process. Two different types of electrolytic cells are used for the process: *diaphragm* or *membrane* cells, in which the reduction and oxidation chambers are separated by a barrier, and *mercury* cells, in which two distinct compartments are used to effect the conversion. Figures 2.6 and 2.7 show the schematics of both the diaphragm/membrane and mercury cells, respectively, with the constitutive reactions shown.

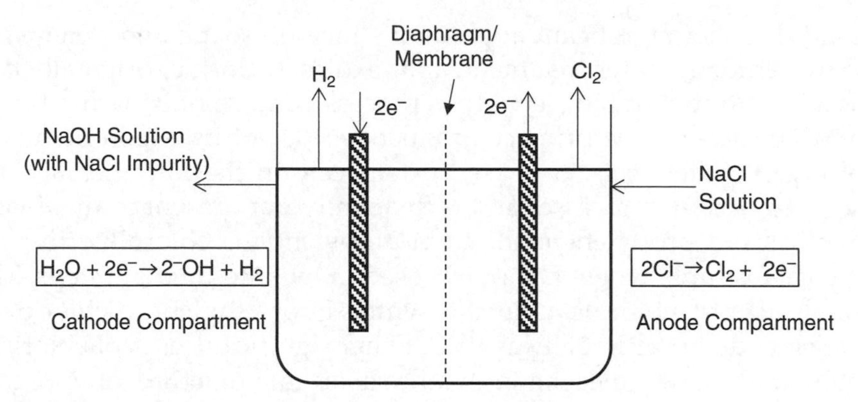

Figure 2.6 Schematic representation of diaphragm/membrane cell used in chloralkali industry.

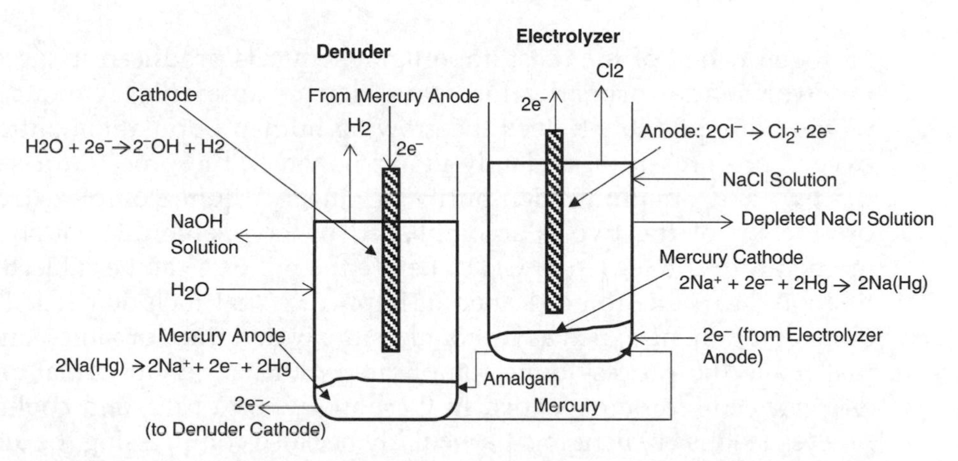

Figure 2.7 Schematic representation of mercury cell used in chloralkali industry.

The oxidation of chlorine in mercury cells is accompanied by the reduction of sodium ion to elemental sodium on a mercury cathode, leading to the formation of sodium amalgam. The amalgam then flows to a denuding chamber where the elemental sodium is oxidized back to sodium ion, with the hydroxide ion formation constituting the reduction reaction. The diaphragm/membrane cells constitute the majority of the electrolyzers, in no small measure due to environmental concerns related to mercury.

Caustic soda is used in a large number of industries, with pulp and paper and organic chemical manufacture accounting for nearly 40% of its use [1]. Other significant applications include soaps and detergents, petroleum refining, water treatment, and textiles. Chlorine, originally considered as a by-product of the electrolysis process because of its limited use, is often now considered the primary product because of its use in the manufacture of organic chemicals. It is used extensively in the intermediate processing steps in organic processes and is frequently not present in the final product. Synthesis of propylene oxide from propylene via chlorohydrin process is a typical example where chlorine is used in the intermediate steps. The largest single use of chlorine is in the synthesis of ethylene dichloride, most of which ends up ultimately as PVC. Other significant uses of chlorine include pulp and paper bleaching, synthesis or manufacture of inorganic compounds, and water treatment.

2.5.3 Nitrogen and Oxygen

Nitrogen is one of the most important chemicals produced in large quantities worldwide, primarily for conversion to ammonia, without which it would be impossible to feed the growing human population. Nitrogen and oxygen are present abundantly in air; however, the process for separating the two and obtaining high-purity products is quite complex. Large-scale production of the two is accomplished by cryogenic distillation of air at temperatures ~83 K (−190°C) [1]. Before the mixture can be subjected to distillation, air needs to be cleaned of impurities that include particulates and other components, such as carbon dioxide, water, trace organics, and others, that make the process quite complicated. Further, a substantial amount of energy needs to be expended in the liquefaction of air, and cooling of the process feed is challenging, particularly because compressing the air for processing generates additional heat. Cooling the air to low temperatures also causes formation of solids—ice and dry ice—that need to be managed and removed properly.

Most of the nitrogen produced is intended for "captive" use; that is, it is produced specifically for manufacturing ammonia [2]. However, there are several other applications of nitrogen, many owing to its nonreactive nature. It provides an inert atmospheric blanket for many processes where the presence of species such as oxygen, water vapor, and others is highly undesirable. Nitrogen is also used for EOR, where it is pumped into oilfield reservoirs to maintain pressure and force oil out for additional recovery.

Pure oxygen, produced along with nitrogen, is valuable for its reactivity. Oxidation is a common process in chemical and other industries, and use of oxygen rather than air has several advantages, including increased efficiencies and higher processing temperatures. Further, significantly lower volumes of gas need to be handled. One of the largest uses of pure oxygen is in the steel industry, where it is used in basic oxygen or open-hearth furnaces. Pure oxygen is also used in the synthesis of ethylene and propylene oxides and in oxyacetylene torches for welding and cutting metals [2].

2.5.4 Hydrogen and Carbon Dioxide

The second element necessary for synthesizing ammonia is hydrogen, by far the most abundant element in the solar system, and on a molar basis, the fourth-most abundant element on earth [7, 8]. However, almost all of this hydrogen is bound in chemical compounds, and molecular hydrogen is obtained from these compounds—hydrocarbons or water. Steam reforming of methane is the dominant production process for hydrogen, accounting for over 48% of the hydrogen produced globally [9]. The fundamental reactions involved in the process are as follows:

$$CH_4 + H_2O \rightarrow CO + 3H_2 \tag{2.1}$$

$$CO + H_2O \rightarrow CO_2 + H_2 \tag{2.2}$$

The reactants and products of both the reactions are in the gaseous phase. Both the reactions are reversible as well; however, only the desired forward reactions are shown in the equations. The first equation represents the steam-reforming reaction, and practically any hydrocarbon can be subjected to this reaction to obtain hydrogen. The second reaction is called the *water-gas shift* reaction, which yields additional hydrogen while also converting the carbon monoxide to carbon dioxide. The gas mixture is then separated, typically through a combination of absorption (of carbon dioxide) and membrane separation processes. Figure 2.8 shows a simplified flowsheet for hydrogen production.

Methane needed for the reforming reactions is obtained from the natural gas, which is subjected to various separation and purification operations to remove the impurities that include sulfur compound and many others. Similar operations are needed when other hydrocarbons are used as raw materials. The steam-reforming reaction is highly *endothermic*, and the heat necessary for the reaction is typically obtained by burning a part of methane. The reaction product—the carbon monoxide–hydrogen mixture—is called

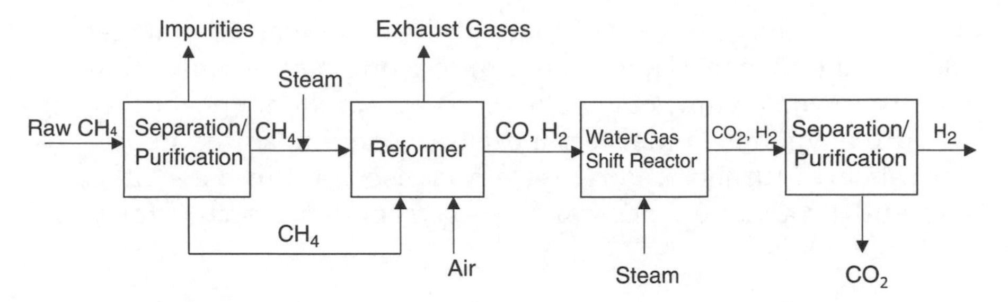

Figure 2.8 Hydrogen production by steam-reforming process.

synthesis gas, or *syngas*, and is subjected to the water-gas shift reaction, typically conducted in two stages. Both the reforming reaction and the water-gas shift reaction are catalytic reactions. The final stage in the process involves complicated separation and purification steps yielding both hydrogen and carbon dioxide.

Invariably, all the hydrogen produced is for captive consumption—either conversion to ammonia or for upgrading hydrogen content and removing impurities in refinery operations. It is also used in the production of other chemical compounds, such as methanol. Hydrogenation is a common reaction step in many organic syntheses. Synthesis gas, a mixture of carbon monoxide and hydrogen at various ratios, can be used for production of transportation fuels using the Fischer-Tropsch process [2].

As can be seen from the figure, carbon dioxide is a by-product of the steam-reforming process. Large quantities of carbon dioxide are also generated through the combustion of the fossil fuels; however, most of this carbon dioxide is generally emitted to atmosphere. Carbon dioxide is used for EOR as well, but its major use is in refrigeration. Other significant uses involve beverage carbonation, maintenance of inert atmosphere, and chemical manufacture [1].

2.5.5 Ammonia

The manufacturing process for ammonia requires highly pure nitrogen and hydrogen gases, which are produced using the two other significant production processes previously described. Cryogenic distillation of air yields the nitrogen necessary for ammonia synthesis, and the hydrogen required is invariably generated by the steam-reforming of methane (natural gas). Although both these processes, in themselves, represent remarkable manifestation of engineering endeavors, the ammonia synthesis process represents a crowning achievement of chemical engineering. The equation for the gas phase reaction is straightforward:

$$0.5\ N_2(g) + 1.5\ H_2(g) = NH_3(g) \tag{2.3}$$

However, the reaction is reversible; that is, product ammonia tends to decompose back into nitrogen and hydrogen. The reaction must be carried out at high enough temperature to have a reasonable rate of reaction. However, increasing the temperature results in limiting the conversion of reactants or formation of ammonia as the reverse reaction starts to become more favorable. Further, nitrogen and hydrogen will react only slowly in absence of a catalyst, and several eminent scientists and engineers spent considerable time, effort, and money on investigating the reaction until a suitable catalyst—iron with various promoters—was found by Fritz Haber, and further engineering development was conducted by Carl Bosch [10]. A simplified block flowsheet of the process for ammonia synthesis is shown in Figure 2.9. A key point to note in this process is that the product gases exiting the converter are recirculated to the reactor after the recovery of ammonia.

As mentioned earlier, the dominant use of ammonia is for making fertilizers: all the N in the fertilizers is ultimately derived from ammonia, through either ammonium ion or nitrate via nitric acid. Ammonia, through nitric acid, also finds applications in explosives. It is also used in the synthesis of polymers, via urea in formaldehyde resins or via various other precursors in nylon.

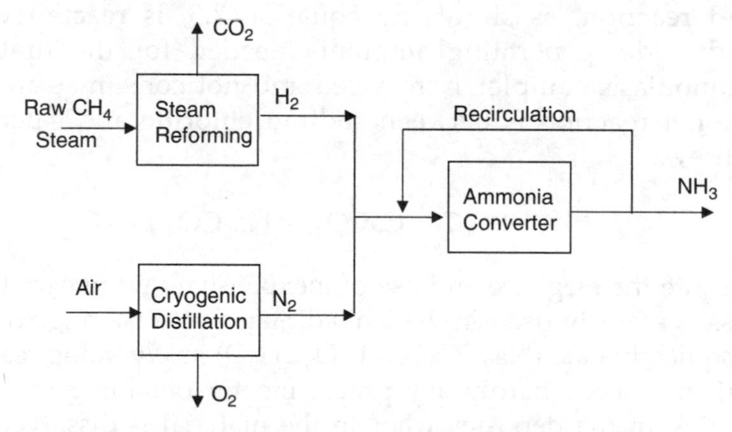

Figure 2.9 Simplified flowsheet of ammonia synthesis process.

2.5.6 Soda Ash (Sodium Carbonate)

Soda ash is one of the largest-volume basic inorganic chemicals, as can be seen from Table 2.4. The biggest use of soda ash is in the glass industry, as nearly 90% of all glass is soda-lime-silica glass [1]. Soda ash, due to its alkaline nature, competes with caustic soda in applications that are related to this alkaline nature. These involve soap/detergent and other chemical manufacture as well as the paper and pulp industry. Soda ash, being a milder alkali than caustic soda, may be preferable in applications that do not require strongly alkaline conditions. Caustic soda is preferred for processes requiring a stronger alkali and for the processes where evolution of carbon dioxide cannot be tolerated or is undesirable.

Historically, soda ash was produced by the *LeBlanc process*, which was replaced by the ammonia-soda or *Solvay process*. The Solvay process is an elegant example of application of chemical engineering principles, with the main reactions as follows [1]:

$$2NH_4OH + 2CO_2 \rightarrow 2NH_4HCO_3 \tag{2.4}$$

$$2NH_4HCO_3 + 2NaCl \rightarrow 2NaHCO_3 + 2NH_4Cl \tag{2.5}$$

$$2NaHCO_3 \rightarrow Na_2CO_3 + CO_2 + H_2O \tag{2.6}$$

The carbon dioxide needed for the reaction is obtained from calcium carbonate decomposition. Calcium oxide formed in the reaction is hydrated to form calcium hydroxide. Ammonium chloride generated in the second reaction, as shown by equation 2.5, is reacted with the calcium hydroxide, generating ammonia needed for the first reaction. Thus, ammonia is completely recycled and not consumed in the process, and the net reaction is between sodium chloride and calcium carbonate as follows:

$$2NaCl + CaCO_3 \rightarrow Na_2CO_3 + CaCl_2 \tag{2.7}$$

Despite the elegance and use of inexpensive raw materials, the Solvay process is scarcely used in the United States because huge deposits of sodium sesquicarbonate ($Na_2CO_3 \cdot NaHCO_3 \cdot 2H_2O$) in Wyoming readily yield a material that needs hardly any processing for obtaining the product. Solution mining of the deposits, wherein the material is dissolved underground in situ and the solution is pumped up, has lowered the cost significantly, making the Solvay process uneconomical.

2.5.7 Ethylene and Propylene

As can be seen from Table 2.4, ethylene is by far the largest-volume basic organic chemical produced. Almost all of the ethylene produced worldwide is derived from crude oil, with some obtained from natural gas. Distillation of crude oil results in a number of different fractions ranging from methane to asphalt and coke. In general, the fractions in the order of increasing carbon number are refinery gases (C1–C4 alkanes), gasoline, naphtha, kerosene, diesel, gas oil, fuel oil, lubricants, waxes, and asphalt and coke [2]. Very little of the light fraction is ethylene, which is produced from ethane, propane, naphtha, and other hydrocarbons through a process called *cracking*. Cracking is a versatile process in refineries whereby molecular bonds in hydrocarbons are cleaved and rearranged to obtain desirable products. Cracking using catalysts, typically in a fluid-bed catalytic cracker, is usually conducted to obtain gasoline [11]. The lighter alkenes, such as ethylene and propylene, are obtained by thermal cracking (also called *steam cracking*) in absence of a catalyst at ~1600°F [1]. Steam cracking of ethane, propane, liquefied petroleum gas, and naphtha yields ethylene. Propylene is also obtained in the process except when the starting material is ethane.

Nearly all of the ethylene produced ultimately ends up in a polymer—directly in low-density and high-density polyethylenes or indirectly through ethylene oxide to glycol and polyesters, through vinyl chloride to PVC, or through styrene to a number of polymeric products. Similarly, a large fraction of propylene ends up as a polymer, either directly as polypropylene or indirectly through propylene oxide and acrylonitrile. Propylene is also converted to intermediates such as cumene and isopropyl alcohol, which are further processed into other chemical products.

2.5.8 Benzene, Toluene, and Xylenes

Toluene does not appear in Table 2.4 yet is included here with the discussion of benzene and xylenes, as all three are present in the aromatic fraction in petroleum refining. Catalytic reforming of naphtha, the process used for obtaining benzene and xylenes, also yields toluene. While a large percentage of toluene is converted to the more valuable benzene through hydrodealkylation, it finds widespread use as organic solvent and raw material for trinitrotoluene and is also converted to toluene diisocyanate for synthesizing polyurethanes. Benzene, combined with ethylene, forms ethylbenzene, which is dehydrogenated to styrene and ultimately converted into polymers. Benzene hydrogenation yields cyclohexane, a solvent and a precursor for nylon 6 and nylon 66. Other significant chemicals obtained from benzene include cumene and aniline, which are further converted into other chemicals, including polymers. The xylene fraction consists of meta (*m*-xylene), ortho (*o*-xylene), and para (*p*-xylenes), in the order of increasing importance. *p*-Xylene is the starting material for terephthalic acid and dimethyl terephthalate. Both the acid and the ester are converted into polyesters, PET and PBT, that find applications in fibers, bottles, automotive components, and many others. *o*-Xylene is generally converted into phthalic anhydride, an intermediate for the synthesis of plasticizers. Most of the *m*-xylene is generally converted into the other two isomers.

In addition to these chemicals, several other high-volume chemicals can readily be recognized from the trade data. The books *Shreve's Chemical Process Industries* [2] and *Survey of Industrial Chemistry* [1] offer far more comprehensive information about a large number of chemicals and should be the first resources to be consulted for such purposes.

2.6 Characteristics of Chemical Industries

The plants and processes in chemical process industries possess the following major characteristics:

- *Large production capacity:* The average production capacities of plants engaged in the manufacture of bulk chemicals such as sulfuric acid, ammonia, and ethylene can easily reach into thousands of tons per day (tpd). The largest sulfuric acid plant in the United States has a capacity to produce 4500 tons of acid per day. A plant producing 100 tpd of sulfuric acid is classified as a small plant. The average size of an ethylene unit is about 2500 tpd. Most chemical processes benefit from the economy of scale; that is, the capital and operating costs do not increase linearly with capacity, making larger units more profitable than smaller ones. The large size of the units also means that the industry is highly capital intensive [1].
- *Continuous operation:* Almost all the plants producing bulk chemicals operate in a continuous, steady-state mode, where conditions are invariant with respect to time. Pharmaceuticals and fine chemicals produced in lower quantities are often produced in batch processes. Even for these plants, while the reaction may be conducted in a batch reactor, the subsequent processing may be carried out in a continuous mode.

- *High productivity:* A consequence of the continuous operation is the high degree of automation in the chemical processes. Increasingly, the processing steps are controlled and run by computers, requiring minimal operator input. The combination of the economy of scale and high degree of automation has resulted in lower manpower requirements, in turn increasing personnel productivity as measured by metrics such as production volume per plant employee.
- *Product diversity:* The industry produces practically an unlimited number of products with widely varying characteristics. The products can be gaseous, liquid, or solid; organic or inorganic; volatile or nonvolatile; acidic, basic, or neutral; water soluble or insoluble; biodegradable or persistent; and many other contrasting properties. The applications also show remarkable extremes: from fertilizers and life-saving drugs to pesticides and defoliants, from cosmetics to solvents that dissolve metals and paints, and so on. The outputs of different processes may not share any characteristics, yet the processes can be amenable to analysis and design based on the same fundamental chemical engineering principles.
- *Energy consumption and integration:* The chemical industry is a major consumer of energy. The Energy Information Administration of the U.S. Department of Energy maintains exhaustive data on all aspects of energy in the United States. According to data from its Manufacturing Energy Consumption Surveys (MECS), the chemical industry accounts for 25% to 30% of the energy consumed by the manufacturing sector. Most chemical processes feature many steps that require energy (heat/electricity) input and many that release thermal energy. Chemical industries seek to manage these energy flows effectively by integrating them across various units.

Chemical industries differ significantly from other manufacturing industries with respect to the following three aspects:

- *Uniform product functionality across scale-up and division:* The output products of the chemical industries retain their physical and chemical characteristics after physical division, right up to the molecular level. Products of other manufacturing industries lose their defining characteristics and functionality if split into parts. For example, machines, engines, tools, and computers cease to be commercial products if cut into parts. The chemical product stream can be split into many divisions or scaled-down to any level without any loss of characteristics as compared to the parent stream.

- *True continuity of operations:* The continuous processes in chemical industries operate in a truly continuous manner, that is, with infinite frequency of product formation, or no interruption in the output stream. Several other manufacturing industries may also operate continuously around the clock—24 hours a day, 365 days a year (except for maintenance). However, the output is invariably in the form of discrete objects that are produced at some frequency with a time-interval between output units.
- *Uniformity of product properties and uses:* The characteristics of a chemical product are essentially independent of the manufacturer and the manufacturing process. For example, acetic acid produced by methanol carbonylation is virtually indistinguishable from that produced by ethylene oxidation. Whether produced by BP Chemicals or Millennium Chemicals, its properties and applications are practically identical. Products of other manufacturing industries retain their identity (manufacturer, model, etc.) and differ to a much greater degree across different manufacturers.

Chemical and allied industries are a mosaic of industries with products that bear no resemblance with each other and yet share many unifying characteristics. With its global reach, the chemical industry offers an opportunity for an engineer in practically any place in the world. The American Chemical Council (ACC, www.americanchemistry.com), a trade association of companies engaged in the chemical business, estimates that the business of chemistry supports 25% of the U.S. gross domestic product. Figure 2.10 shows the

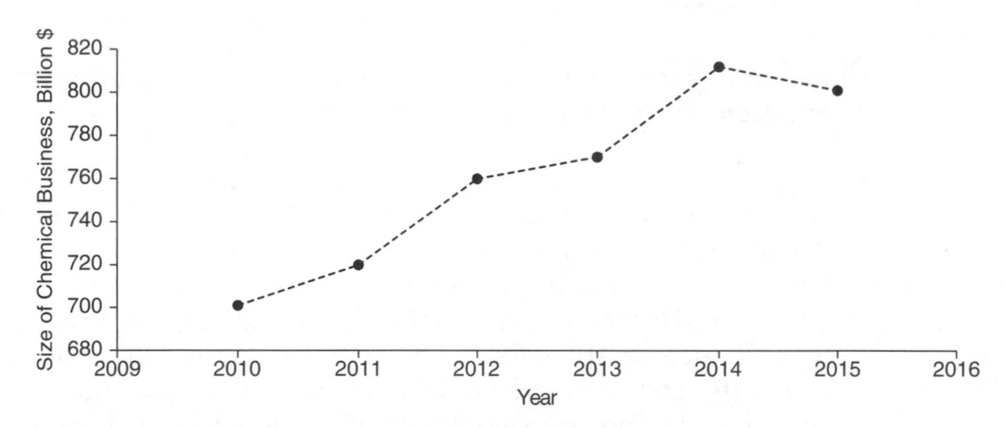

Figure 2.10 Growth of chemical business from 2010 to 2015.

growth in the chemical business over the past 6 years, with its 2015 contribution being $801 billion, reflecting an average annual growth rate of ~2.5% since 2010. According to the ACC, each job in the chemical sector generated nearly 6.3 other jobs in the economy [12].

Even the briefest of introduction to just a few of the tens of thousands of chemical products and the chemical processes for manufacture should make clear the wide variety of challenges faced by a chemical engineer. The education of a chemical engineer must prepare him/her to deal with any type of reaction or separation and handle any chemical while contributing to the profitability of the enterprise within the legal, ethical, and moral constraints placed by the society. How this is accomplished by the chemical engineering programs is described in the next chapter.

2.7 Summary

Chemical and allied industries play an important role in a nation's economy and are a major constituent of the manufacturing sector. Although the largest fraction of the output of chemical industry goes as a feedstock to other industries, the industry also produces many consumer products vital to society. The global nature of the chemical industry can be understood from the largest chemical companies in the world. The distinctive characteristics of the chemical industry offer exciting and challenging opportunities for the engineering professional.

References

1. Chenier, P. J., *Survey of Industrial Chemistry*, Second Revised Edition, VCH, Weinheim, Germany, 1992.
2. Austin, G. T., *Shreve's Chemical Process Industries*, Fifth Edition, McGraw-Hill, New York, 1984.
3. Tullo, A. H., "C&EN's Global Top 50," *C&E News*, Vol. 94, No. 30, 2016, pp. 32–37.
4. Tullo, A. H., "Top 50 U.S. Chemical Producers," *C&E News*, Vol. 94, No. 19, 2016, pp. 16–19.
5. Office of Energy Efficiency and Renewable Energy (EERE), "Chemicals Industry Profile," http://energy.gov/eere/amo/chemicals-industry-profile, webpage accessed August 26, 2015.

6. Brueske, S., C. Kramer, and A. Fisher, "Bandwidth Study on Energy Use and Potential Energy Saving Opportunities in U.S. Chemical Manufacturing," Office of Energy Efficiency and Renewable Energy (EERE) Report, June 2015.
7. Emsley, J., *The Elements*, Third Edition, Oxford University Press, Oxford, England, U.K., 2000.
8. Stewart, P. J., "Abundance of the elements—A new look," *Education in Chemistry*, Vol. 50, No. 1, 2003, pp. 23–24.
9. Bhat, S. A., and J. Sadhukhan, "Process intensification aspects for steam methane reforming: An overview," *AIChE Journal*, Vol. 55, No. 2, 2009, pp. 408–422.
10. Ertl, G., "The Arduous Way to the Haber-Bosch Process," *ZAAC*, 2012, pp. 487–489.
11. Vogt, E. T. C., and B. M. Weckhuysen, "Fluid catalytic cracking: Recent development in the grand old lady of zeolite catalysis," *Chemical Society Reviews*, Vol. 44, 2015, pp. 7342–7370.
12. American Chemistry Council (ACC), *2015 Guide to the Business of Chemistry*, www.americanchemistry.com.

Problems

2.1 Discuss the role of the chemical industry in society, focusing on socioeconomic aspects.

2.2 What are some of the important chemicals categorized as "other chemical products" (section 2.2.7)?

2.3 Aspirin is one of the most important pharmaceutical products produced in large quantities. What are the raw materials needed? How does the manufacturing process for aspirin differ from that for sulfuric acid?

2.4 Select any two companies from those listed in Tables 2.2 and 2.3. What are some of the important products produced by these companies? Are these intermediates or consumer products?

2.5 Prepare summaries describing the manufacturing process and applications for the important, high-volume chemicals not explicitly discussed in this chapter: calcium hydroxide, phosphoric acid, nitric acid, methane, acetylene, and butadiene.

2.6 Syngas is an important commodity that is valuable both as fuel and as a feedstock for synthesizing chemicals. What are some of the important applications of syngas? Examine the Fischer-Tropsch process.

2.7 What are some of the important polymer products (other than polyolefins and polyesters)? What are the raw materials (monomers) for synthesis of these polymers (polyamides, polyurethanes, etc.)? What types of reactions do they undergo?

2.8 Select any five important chemical products. Collect the information related to major manufacturers, production volume, manufacturing technology, typical plant capacities, uses, and any other aspect to build a comprehensive database.

2.9 Discuss the similarities and differences between the chemical industries and other manufacturing industries.

2.10 How does the energy consumption of the chemical sector compare to that of the rest of the economy? Which chemical products account for the largest share of consumed energy?

CHAPTER 3

Making of a Chemical Engineer

Only when I began studying chemical engineering at Oregon Agricultural College did I realize that I myself might discover something new about the nature of the world.

—Linus Pauling[1]

Evolution of the engineering profession and the establishment of chemical engineering as a separate and distinct engineering discipline were described in Chapter 1, "The Chemical Engineering Profession." The vast breadth of career options for chemical engineers in the chemical and allied industries was presented in Chapter 2, "Chemical and Allied Industries." It should be apparent that teaching an individual to transform into a professional who can potentially be employed productively in any of the chemical and allied industries described in the previous chapter is a significant challenge. Remarkably, the colleges and universities engaged in chemical engineering education manage to accomplish this challenging task, converting high school graduates into chemical engineers in just 4 years. To understand how this is achieved, it is worthwhile to first take a look at a typical chemical process.

3.1 A Chemical Process Plant: Synthesis of Ammonia

Synthesis of ammonia from nitrogen and hydrogen is often considered the most important chemical process for mankind, and rightfully so [1]. The element nitrogen is involved in every level of biological function in all living organisms. Despite the inexhaustible reservoir of diatomic molecular nitrogen in the atmosphere, the major limitation to growth of living beings is the availability of elemental nitrogen for basic biochemical processes within a living being, as very few of the living

1. Widely considered to be one of the greatest scientists in history, Linus Pauling is a double Nobel laureate with prizes for chemistry in 1954 and peace in 1962. Quotation source: Marinacci, B., *Linus Pauling in His Own Words*, Simon and Schuster, New York, 1995.

organisms have the ability to fix molecular nitrogen. Bacteria belonging to the family *Rhizobia* are the most important microorganisms that are able to fix nitrogen by converting it to ammonia at ambient conditions. Growth of agriculture, and hence the food supply, was limited in the past by the availability of such bacteria to replenish the soil nitrogen used by crops. Industrial synthesis of ammonia allowed mankind to overcome this limitation, making it possible to feed and sustain the billions of human beings who now inhabit the earth [2].

For a long time, it was considered impossible to effect a synthetic reaction between nitrogen and hydrogen at a rate that would make the process amenable to commercial exploitation. Several eminent personalities, including Ostwald[2] and Nernst,[3] conducted detailed investigations without much success. Ultimately, Fritz Haber, a German chemist, demonstrated in his laboratory that it was possible to obtain reasonable yields of ammonia through the use of iron-nickel catalyst [3]. This discovery earned Haber the Nobel Prize in Chemistry in 1918.[4] The commercialization of the reaction for bulk production of ammonia was accomplished with the involvement of Carl Bosch, leading to the Haber-Bosch process, which is the foundation for fertilizer production worldwide. Bosch was awarded the Nobel Prize in Chemistry in 1931 (along with Friedrich Bergius) for his pioneering work in high-pressure chemistry [4].

An abbreviated flowsheet[5] of the ammonia production process is shown in Figure 3.1 [5]. Nitrogen-hydrogen mixture is fed continuously to the ammonia converter, which is a catalytic reactor where the ammonia synthesis reaction takes place. The product stream consisting of ammonia and unreacted reactants is continually withdrawn from the reactor. This product stream exiting the reactor is not useful in its form, as the subsequent ammonia-based processes (nitric acid and fertilizer production, etc.) require pure ammonia. Further, the unreacted reactants are also valuable and need

2. Wilhelm Ostwald received the Nobel Prize in Chemistry in 1909 for contributions in the fields of catalysis, equilibrium, and kinetics.

3. Walther Nernst received the Nobel Prize in Chemistry in 1920 for contributions in the fields of thermodynamics and electrochemistry.

4. An unintended and unfortunate consequence of this ability to synthesize ammonia was that Germany was able to synthesize ammonium nitrate needed for explosives and ammunition, a development that prolonged World War I (Max Perutz, *I Wish I Had Made You Angry Earlier, Essays on Science, Scientists and Humanity*, Cold Spring Harbor Laboratory Press, Cold Spring Harbor, New York, 2003).

5. A flowsheet is a schematic representation of the process, showing various pieces of equipment used in the process and the material flows in and out of these pieces of equipment.

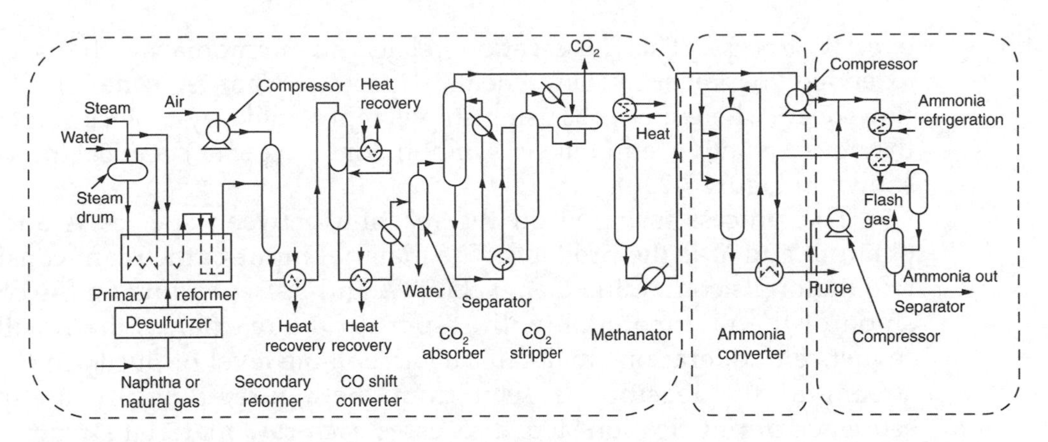

Figure 3.1 Ammonia synthesis process plant.
Source: Austin, G. F., *Shreve's Chemical Process Industries*, Fifth Edition, McGraw-Hill, New York, 1984.

to be recovered. The product separations section accomplishes both the objectives—yielding pure ammonia product, and recovering the unreacted reactants for recycling to the reactor.

Just as processes utilizing ammonia require pure ammonia, the ammonia synthesis reaction requires pure nitrogen and hydrogen reactants. The processes shown upstream of the ammonia converter are required for just this purpose. Neither nitrogen nor hydrogen, while abundant, occurs naturally in pure form. Nitrogen is present as a molecular species, but mixed with oxygen in air. Hydrogen is inevitably found in a bound form—in compounds with oxygen and carbon. The predominant process for obtaining hydrogen is steam reforming of hydrocarbons (primarily methane). Essentially, carbon-hydrogen bonds in the hydrocarbon are cleaved by forming carbon monoxide and then carbon dioxide. The energy required for the process is obtained by the combustion of the hydrocarbon. Oxygen needed for the combustion is obtained from air, consumption of which generates nitrogen. Further separations are conducted to remove all other components, ultimately yielding the desired reactant stream, which can be fed to the ammonia converter.

The various processes and operations conducted in the ammonia plant can be grouped into three blocks, as shown in the figure: raw material processing, ammonia reaction, and product separation. The ammonia converter—the reactor—is the heart of the process. However, several other operations play equally vital roles to make the process functional. As can be seen from this flowsheet, the reactor itself is only a small component in the overall process. This observation related to ammonia synthesis can be extended to practically any chemical process plant. A generalized block flowsheet (where any processing step is represented by a block rather than the representative equipment symbol) for a typical chemical process is shown in Figure 3.2 [6].

The process is based on the reaction between species A and B to obtain X and Y as the products. The material input to the plant consists of raw materials containing the reactants A and B. Depending on the composition of the raw material (with respect to the reactants), there will be a sequence of operations to obtain the acceptable level of purity in the feed stream to the reactor. As seen from the figure, typically a separate sequence of purification steps processes each raw material stream, yielding a pure reactant that is fed to the reactor. The reactants of desired purity are fed to the reactor and subjected to appropriate processing conditions—temperature, pressure, reaction time, and so on—to be converted into the products. A desired reaction is invariably accompanied by a number of side reactions. Further, some fraction of the reactants is almost always left

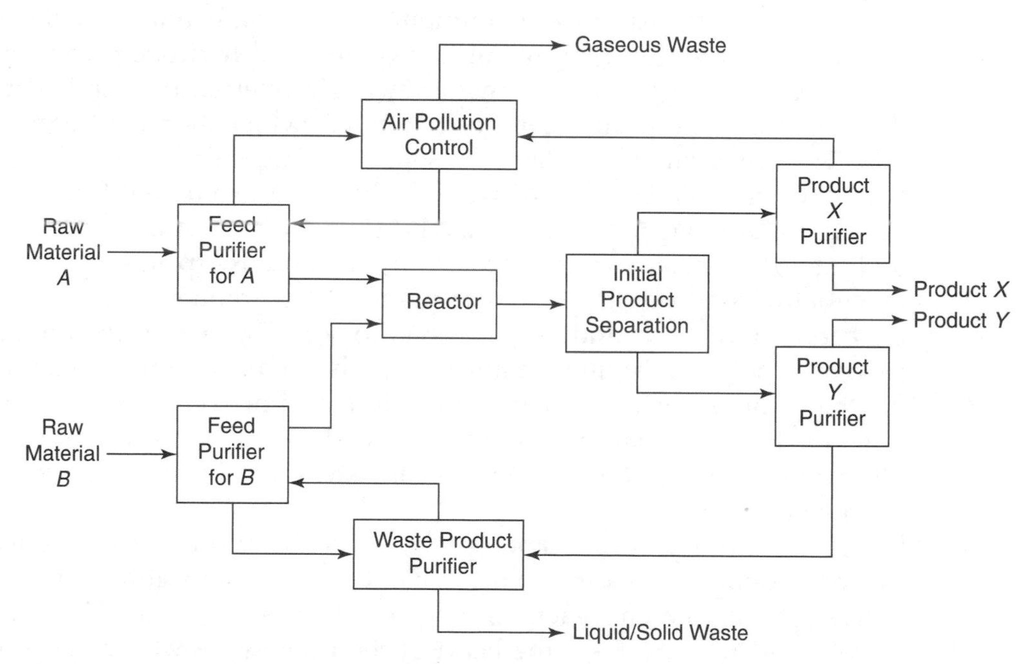

Figure 3.2 Generalized block flowsheet of a chemical process plant.
Source: Luyben, W. L., and L. A. Wenzel, *Chemical Process Analysis: Mass and Energy Balances*, Prentice Hall, Englewood Cliffs, New Jersey, 1988.

over in the reactions. The reactor outlet stream—the raw product stream—consists of the desired products, unconverted reactants, and products resulting from the side reactions. This raw product stream is then subjected to another sequence of operations to separate the product streams. If the raw product stream contains significant quantities of unconverted reactants, those may be separated and recycled back to the reaction stage. Streams devoid of any commercial value are discharged to the environment after proper treatment to ensure compliance with the relevant environmental regulations put in place for the protection of human health and environment.

An individual employed as a chemical engineer in such a process should be able to handle responsibilities associated with each step in the process. He/she should have sufficient knowledge to tackle challenges to ensure smooth operation of the plant from the beginning of the raw material processing to final product formulation and disposition of waste streams. These functions and responsibilities, based on the various technical aspects of materials processing, are described in the following section.

3.2 Responsibilities and Functions of a Chemical Engineer

The major responsibilities and functions of a chemical engineer in a chemical process plant can be identified on the basis of the previous discussion:

- *Design and operation of the reactor:* The chemical reactor is the key unit of the process where the raw materials are converted into desired products. The success of the process is dependent on efficient, economic functioning of the reactor. A chemical engineer should be able to obtain the fundamental information about the intrinsic rate of reaction and its dependence on system parameters, and on the basis of this information, specify the type and size of the vessel that would be used to conduct the reaction. He/she should be able to understand the effect of operating parameters on the reaction and manipulate the conditions to obtain the desired production rate and product composition and purity.
- *Design and operation of separation equipment:* The vast majority of units in the process plant are aimed at accomplishing physical separation of components: obtaining reactants of the desired purity from the available raw materials, recovering the desired product from the product stream exiting the reactor, and removing pollutants from waste

streams before they are released into the environment. In general, no chemical reaction or changes in molecular species occur in the separation units. However, reactive separations, where separations are effected by incorporating a chemical reaction in the separation scheme, are not uncommon. A chemical engineer should be able to quantify the nature of interactions among various components of any stream, identify the separation technique and obtain fundamental thermodynamic and kinetic data, and design the equipment to accomplish any separation needed in the process. He/she should be able to manipulate the operational parameters to obtain the desired purity in the separated streams.

- *Design and operation of material and energy transfer equipment:* Even a cursory glance at a chemical plant will reveal a number of units interconnected with an impressive network of pipes. This network is needed to transfer vast quantities of material through various processing steps. A chemical engineer should be able to design and operate an efficient system to accomplish these transfers. This system will typically include pumps, compressors, and piping for transferring fluids—gases and liquids. Transfers of solid materials can be effected by dissolving or suspending them in appropriate liquids and pumping the solutions or slurries. Solids can also be conveyed pneumatically using gases or on conveyer belts. Chemical process plants also involve transferring vast quantities of energy required for separations and conducting reactions. A chemical engineer should also be able to design systems for transferring energy to and from process streams to achieve and maintain process units at their desired design temperatures. As mentioned in Chapter 2, the chemical sector is one of the highest energy-consuming sectors of the economy. Most chemical processes are energy intensive, and a chemical engineer should be able to minimize the energy cost by properly designing and operating the energy transfer equipment.
- *Process control:* All the steps in a chemical process—whether physical separations or chemical reactions—are typically designed to occur at specific conditions. Deviation from these design operating conditions inevitably results in formation of off-quality product or incomplete separations of components. Maintaining the process plant conditions (pressure, temperature, flow rates, concentrations, etc.) at their setpoints is absolutely critical for meeting product quality specifications and efficient operation of the plant. A chemical engineer must be able to design and tune the control system to ensure that the process operates as

designed and corrective actions are taken to counteract any disturbances within the processing units. These corrective actions should allow the process unit to return to its setpoint within a reasonable period of time such that the disturbances are not propagated throughout the plant.

Each chemical process is unique with respect to the species involved, reactions, and separations; however, analysis of each step is based on unifying scientific and engineering principles. The education of a chemical engineer does not involve teaching specifics of a particular process but rather imparting a knowledge of these unifying principles and concepts that can be applied to any process. The curriculum that accomplishes this educational objective is described in the next section.

3.3 Chemical Engineering Curriculum

The essential components of a chemical engineering curriculum are presented in the sections that follow. The courses corresponding to each one of the four responsibilities are described first, followed by the description of engineering science and specialized fundamental chemical engineering courses and the basic science and mathematics courses that serve as the foundation for engineering education. The role and contribution of general education courses are also outlined.

3.3.1 Advanced Chemical Engineering Courses

The advanced chemical engineering courses are typically taught in the third and fourth (junior and senior) years of the program. Students are exposed to the concepts underlying reactor design, separations, transport processes, and process control. In addition to these courses, the students generally take several technical elective courses based on their interests. A final capstone design course that requires the students to synthesize the concepts from all courses is an essential requirement for all chemical engineering programs.

3.3.1.1 Chemical Reaction Engineering

The course listed most often as *Chemical Reaction Engineering* or *Chemical Engineering Kinetics* explains and teaches the design of chemical reactors. The essence of the design problem is explained simply as follows: *It is desired to produce a certain quantity of a chemical determined by the market demand for that chemical. The engineer is required to specify the size of the vessel and the operating conditions for producing the chemical at the desired rate.*

To solve this design problem, the engineer needs to understand the factors governing the *rate of the reaction*, which is generally defined in terms of the quantity reacting per unit time per unit measure of reactor size. The *intrinsic kinetics* of the reaction is dependent on the concentrations of species involved in the reaction as well as the temperature. Additional factors that play a role in determining the actual or observed rate in the reactor may include flow rate and pattern, agitation speed, and so on.

The course on reactor and reaction engineering teaches students the fundamental concepts and techniques related to reactor design. The desired outcome of the course is competency on the part of the student in design and analysis of chemical reactors for the different types of reactors employed in chemical processes. Following are some of the core concepts taught in the course:

- *Determination of intrinsic kinetics:* The student learns about quantifying the rate of reaction and factors determining this rate. The parameters in the rate expression (equation representing the relationship between the rate and factors affecting the rate) need to be determined experimentally, and the student learns about conducting laboratory experiments for this purpose, as well as analysis and interpretation of the data obtained.
- *Design of reactors:* The types of reactors used in the chemical industry can be categorized on the basis of a number of different criteria: whether they are operated in a batch mode, where conditions change with time, or in a continuous mode, where conditions are invariant and the system is at a steady state; whether the reactors operate under isothermal (constant temperature) or nonisothermal (temperature varying with time or position within the reactor) conditions; whether the reaction is homogeneous (single phase) or heterogeneous (two or more phases); and so on. Figure 3.3 shows the schematic sketches of batch and continuous reactors used in the chemical industry [7]. The operational characteristics of batch and continuous reactors are qualitatively and quantitatively different from each other. Design of batch reactors involves determining the batch size and time, whereas the design of continuous reactors involves determining the reactor size or residence time. Continuous reactors are further classified according to

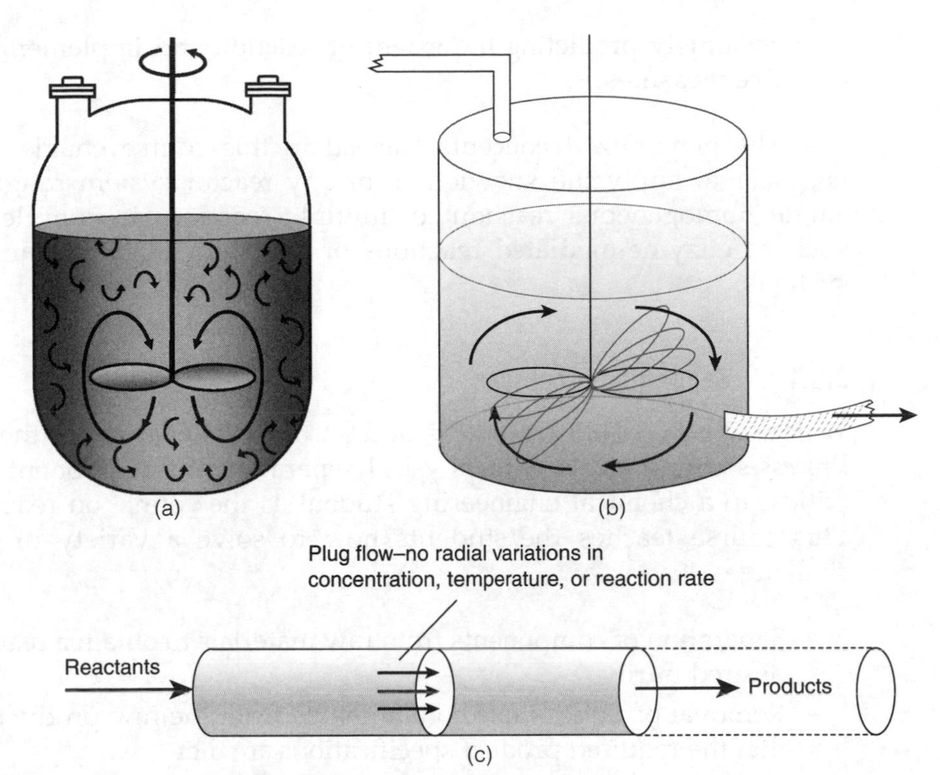

Figure 3.3 Batch and continuous reactors—(a) batch reactor, (b) continuous stirred tank reactor, (c) tubular plug flow reactor.
Source: Fogler, F. S., *Elements of Chemical Reaction Engineering*, Fourth Edition, Prentice Hall, Upper Saddle River, New Jersey, 2004.

the flow pattern. Design equations for mixed flow reactors (contents well mixed) are different from segregated flow reactors (contents not intermixed). A large number of industrial reactions are fluid-solid heterogeneous reactions, the solid being a catalyst. Many other reactions involving a solid phase are noncatalytic; that is, the solid participates in the reaction. Each of these cases requires a distinct approach for reactor design, which a chemical engineering student must understand and master.

- *Analysis of flow characteristics:* The mixed flow reactor and the segregated flow reactor represent ideal flow reactors. The actual flow pattern in operating reactors frequently does not conform to the idealized flow pattern. A student must learn to model the flow pattern and develop analytical techniques to identify the departure from ideality for accurately predicting the extent of reaction and implementing corrective measures.

The generalized concepts learned in this course enable a chemical engineer to apply the knowledge for any reactor system ranging from a single homogeneous reaction to multiple reactions in complex systems, such as enzyme-mediated reactions of complex substances in biological systems.

3.3.1.2 Separation Processes

Variously called *Unit Operations* or *Mass Transfer Operations*, the *Separation Processes* course deals with physical separations of components and is as critical to a chemical engineering student as the course on reactor design. This course teaches the students how to solve a variety of separation challenges:

- Separation of components from raw materials to obtain a reactor feed of desired purity
- Removal of undesirable contaminants from the raw product stream so that the required product specifications are met
- Recovery of valuable components from the waste stream before it is released to the environment

A student, after this course (or courses) on separations, will be able to make a decision regarding the type of separation process and design the equipment to accomplish the desired separation. Following are the most common separation techniques employed in the chemical industry:

- *Distillation*: Distillation is the most common technique utilized by the chemical industry for separating miscible components that exist in liquid phase at the given conditions. Distillation is based on differing volatilities (ease of vaporization) of the components of the mixture. For example, a distillation column used in petroleum refining allows us to separate gasoline from diesel (and aviation fuel and other components). Similarly, ethanol-water separation can be effected by distillation. Figure 3.4 shows a general schematic of a distillation column [8].

  Feed F to the column is separated into distillate D and bottoms B. The energy needed for the separation is supplied in the reboiler at the

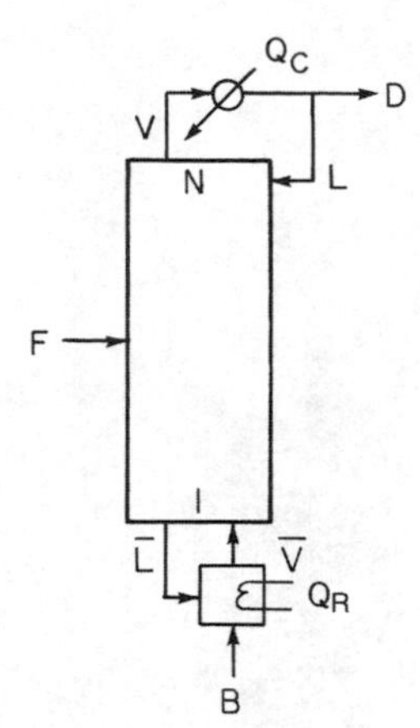

Figure 3.4 Schematic of a distillation column.
Source: Wankat, P. C., *Separation Process Engineering*, Third Edition, Prentice Hall, Upper Saddle River, New Jersey, 2012.

bottom of the column to generate a vapor stream. This vapor stream is contacted with a liquid stream flowing down the column. A condenser located at the top of the column condenses the vapor at the top, returning part of the condensed liquid back to the column. The distillate product at the top is preferentially enriched in the more volatile components of the feed, whereas the bottom product is enriched in the less volatile components. Figure 3.5 shows a system of three distillation columns used for solvent recovery in a chemical plant.

- *Gas absorption/stripping:* Both gas absorption and stripping operations involve transferring a component from one phase to another. Removing a contaminant gas (such as H_2S or sulfur oxides) from a gas stream (prior to venting the stream to the atmosphere) by absorbing it in an alkaline solution is an example of gas absorption. Stripping is the reverse of absorption wherein a dissolved gas is stripped from a liquid. These operations are based on manipulating the affinity of the target component for one phase over the other. Figure 3.6 shows a schematic of an absorber-stripper system.

The gas to be treated for the removal of a component is fed at the bottom of the absorber, where it is contacted with a solvent that preferentially dissolves the component. Solvent needed for absorption is regenerated in the stripping column, where the solvent exiting the absorption column that is enriched in the component is contacted with the stripping gas. The stripping column operates at a higher temperature; as a result, the component dissolved in the solvent is transferred to the gas phase.

Figure 3.5 A chemical plant with three distillation columns.
Source: Wankat, P. C., *Separation Process Engineering*, Third Edition, Prentice Hall, Upper Saddle River, New Jersey, 2012.

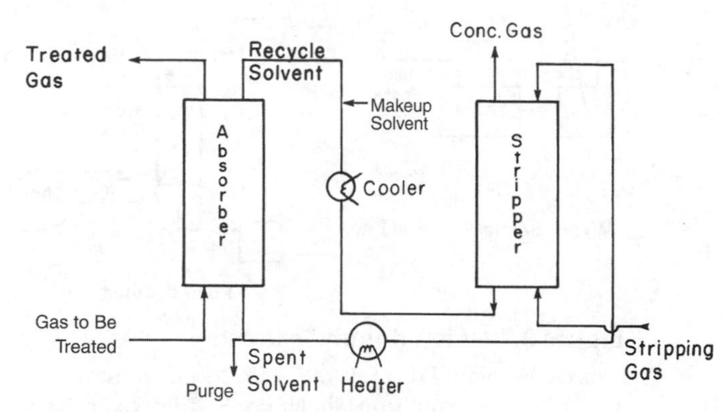

Figure 3.6 Absorber-stripper system.
Source: Wankat, P. C., *Separation Process Engineering*, Third Edition, Prentice Hall, Upper Saddle River, New Jersey, 2012.

- *Liquid-liquid extraction:* A liquid stream is often purified by transferring the impurity to another liquid stream that is immiscible with the feed stream. As with absorption/stripping, manipulating operating conditions allows an engineer to change the affinity of the component of interest. Typically, but not always, one liquid stream is aqueous while the other one is organic. Figure 3.7 shows a conceptual schematic of separation using liquid-liquid extraction using two different contacting devices.

  In the mixer-settler unit shown on the left, the two immiscible phases are stirred together to promote mixing and transfer of the component of interest. The two phases are separated in the settler unit under the influence of gravity, taking advantage of the density differences. The light and dense phases are removed from different locations in the settler unit. The mixing of phases and the resultant transfer of the component is accomplished by providing a contact surface by means of a packing or stages in a pulsed column. Pulsing the column results in regeneration of surface, promoting the transfer of the component. The phases are separated, as shown in the figure.
- *Adsorption:* A component is removed from the fluid (gas or liquid) phase by contacting the fluid with a solid. The component is preferentially sorbed to the surface of the solid and essentially removed from the fluid. The principle of the pressure-swing adsorption cycle used for clean-up of an impurity containing gas is shown in Figure 3.8.

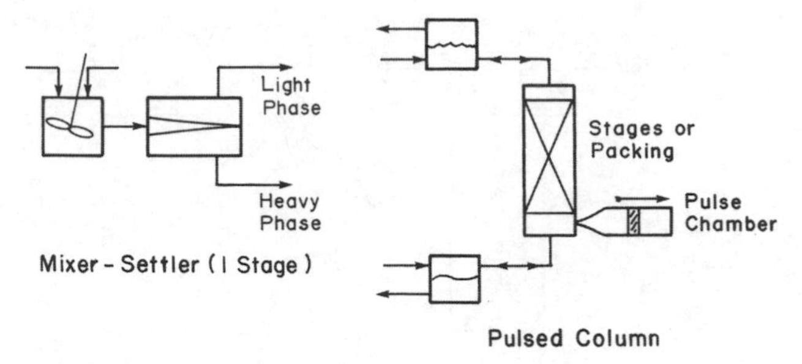

Figure 3.7 Liquid-liquid extraction.
Source: Wankat, P. C., *Separation Process Engineering*, Third Edition, Prentice Hall, Upper Saddle River, New Jersey, 2012.

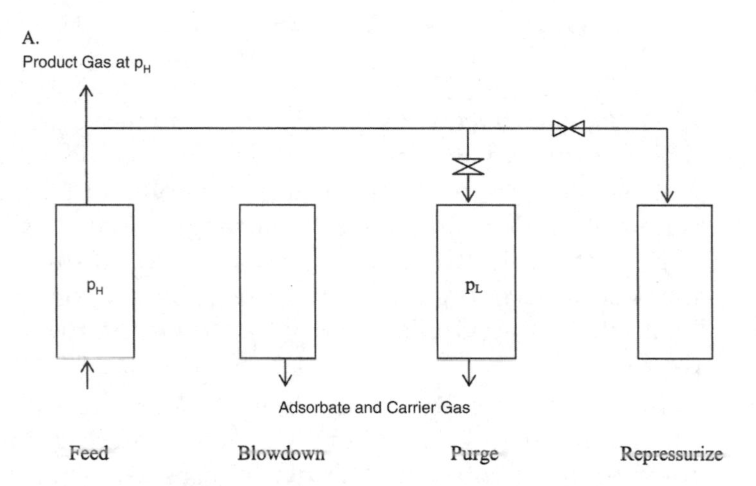

Figure 3.8 Principle of pressure-swing adsorption.
Source: Wankat, P. C., *Separation Process Engineering*, Third Edition, Prentice Hall, Upper Saddle River, New Jersey, 2012.

The gas containing the impurity is contacted with the adsorbent in a column at high pressure. The adsorbent preferentially sorbs the impurity removing it from the gas, which is collected at the product gas at high pressures (P_H). The adsorption column operates in a batch mode; that is, the adsorbent gets progressively saturated with the impurity, losing the ability to clean the gas. At this point, the feed gas flow is stopped, and the column pressure lowered. The third stage consists of purging the column at this low pressure (P_L), where the impurity picked by the adsorbent is desorbed or removed, regenerating the column capacity. The last stage is repressurizing the column to high pressure in order to repeat the cycle.

- *Evaporation and drying:* A stream dilute with respect to a solute is often subjected to evaporation for concentration. For example, a dilute caustic solution is the first product in the caustic soda (NaOH) manufacture. Progressively concentrated solutions (50%, 70%, etc.) are obtained through evaporation. Drying involves continuing this solvent removal to its extreme limit, leaving behind only the pure solute. The solvent can be aqueous (as in the majority of these operations) or organic. These operations consume significant amounts of energy, and an improperly designed evaporator or dryer can impact the plant economics adversely.
- *Membrane separations:* Figure 3.9 illustrates the principle of separations using membranes. Membranes are semipermeable barriers between two phases, and the separation of a mixture is based on the preferential transfer of one or more of the components across this barrier.

  The schematic on the left shows a crossflow unit where the component transferring across the membrane is swept from the system using a carrier gas. The components not diffusing across exit the unit as retentate. The schematic on the right in Figure 3.9 shows a hollow fiber membrane unit, where permeate diffuses across the tube walls from the feed-retentate stream flowing inside the tubes. Important applications of membrane separations include oxygen-nitrogen separation and desalination.

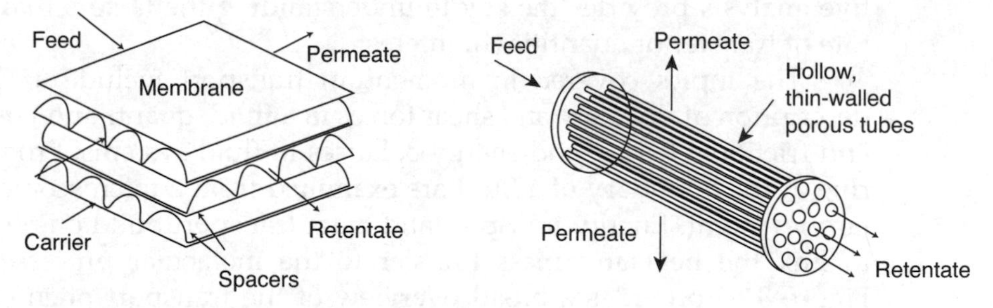

Figure 3.9 Separation using membranes.
Source: Wankat, P. C., *Separation Process Engineering*, Third Edition, Prentice Hall, Upper Saddle River, New Jersey, 2012.

Several other important separation processes—ion exchange, electrochemical separations, crystallization, and so on—are also studied in this course. The generalized concepts covered in the course provide the foundation for a chemical engineer to design and operate a separation scheme for any mixture.

3.3.1.3 Transport Phenomena

Almost all chemical engineering curricula feature a sequence of two (if on a semester system) or three (if on a quarter system) courses involving the term *transport* in the course name. These courses are typically offered in the junior (or third) year of the program and are termed *Transport and Rate Processes* at the University of Idaho. *Transport Phenomena* is a common name used in many other institutions. The three transport phenomena covered in these courses are momentum transport, energy transport, and mass transport [9].

These courses lay the theoretical foundation for understanding the processes occurring in chemical systems. All processes, whether they involve a simple fluid flow, transfer of heat, or a component, occur at the molecular level. The mechanisms of transfer of the quantities involved—momentum, energy, and mass—are analogous to each other. The theoretical analysis and quantitative results obtained from examining the molecular-level processes for momentum transfer can be extended and applied to the other two transport phenomena.

These courses typically begin with the mathematical description of phenomena associated with the flow of fluids. The similarity between fluid flow behavior and energy/mass transfer is used to extend the mathematical model of momentum transport to energy and mass transport. This quantitative analysis provides the key to understanding the factors that govern the rate of transfer of quantities of interest.

The topics covered in momentum transport include a fundamental description of viscosity and shear forces in a fluid, quantitation of turbulence and frictional losses, and energy balances in fluid systems. Processes occurring at the boundary of a fluid are examined from a microscopic or molecular viewpoint. Energy transport and mass transport build on these concepts to link the heat and mass transfer to the molecular processes in fluids. Figure 3.10 provides a broad overview of the transport phenomena topics and some important concepts.

The system behavior at the macroscopic level can always be linked to measurable parameters, such as temperature, flow rate, and so on, using a

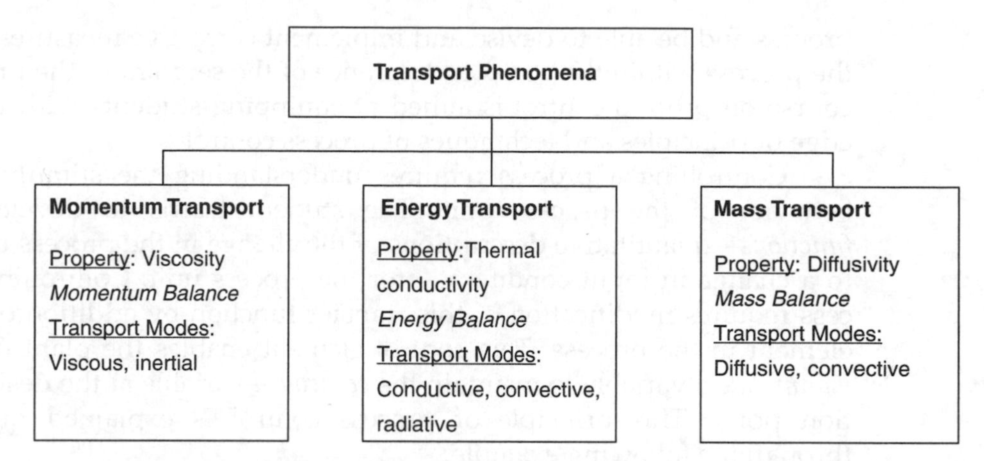

Figure 3.10 Overview of transport phenomena.

trial-and-error procedure. Empirical relationships obtained through such an exercise have a limited utility and validity for specific situations. Study of transport phenomena occurring at the microscopic level provides a theoretical basis for explaining the system behavior at the macroscopic level. Relationships between observed quantities and system parameters developed on the basis of this theoretical analysis are scientifically valid and have general applicability to other systems. Transport phenomena courses thus equip a student with the knowledge to analyze any situation based on sound science. This knowledge imparts an ability to predict accurately the system response to changes in operating parameters and hence an ability to increase the efficiency of operations, whether it is separations, heat transfer, or simply material flow. The quantitative tools acquired by the student are used in the design of heat transfer, fluid flow, and separations equipment.

3.3.1.4 Process Control

Most process plants operate in continuous, steady-state mode. Maintaining the plant conditions at the specified design values is essential for meeting the product specifications. At the same time, no processing step is immune from experiencing disturbances due to variations in the quality of raw material streams, malfunctioning of units (pump failure, loss of electrical power, etc.), inlet flow rate fluctuations, and many other factors. A chemical engineer must understand the impact of such disturbances on the output of the process and be able to devise and implement corrective measures to control the process within the accepted tolerance of the setpoint of the process. The course on process control is aimed at equipping students with the knowledge of principles and techniques of process control.

Controlling a process requires understanding the stimulus-response behavior of the process unit. The student learns to develop *transfer functions*—quantitative descriptions of the change in the process output due to a change in input conditions—for the process unit. Controlling the process requires modification of this transfer function by addition of a control element in the process. This control element enables the plant operator to *manipulate* a variable to maintain the *controlled* variable at the desired operation point. This principle of process control is explained qualitatively through the following example.

Consider a holding tank that has a continuous inflow and an outflow of a liquid stream. It is required to control the liquid level in the tank at a set value. It is clear that any variation in the inlet stream flow rate, if not controlled, will cause the liquid level to change. When the inlet flow increases, the control strategy should result in increased outflow. The opposite must occur when the inlet flow decreases. This can be accomplished by installing in the outlet line a valve that can be opened or closed on the basis of the disturbance. The tank level is the *controlled* variable, the desired level is the *setpoint*, and the outlet flow (more accurately, the valve position) is the *manipulated* variable. The control scheme is shown in Figure 3.11 [10].

The transfer function for the process describing the relationship between the controlled variable and the feed (inlet flow) is developed, and the desired change in the manipulated variable is quantified on the

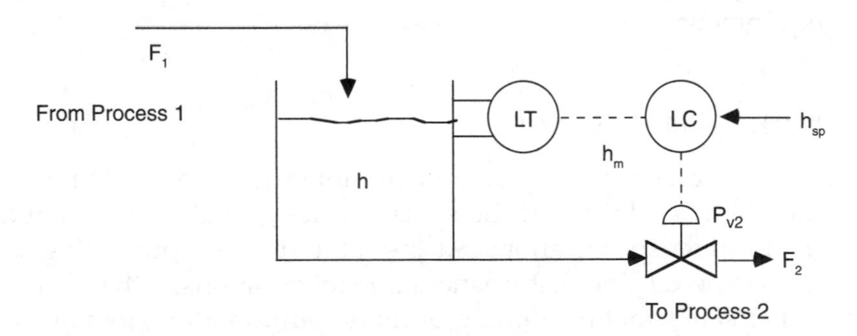

Figure 3.11 Controlling liquid level in a process tank.
Source: Bequette, B. W., *Process Control: Modeling, Design and Simulation*, Prentice Hall, Upper Saddle River, New Jersey, 2003.

basis of this transfer function. The process is monitored by making certain measurements. The *measured* variable may or may not be the same as the controlled variable. In this case, the liquid level or inlet flow rate may serve as the measured variable. It can be seen that several alternative strategies are possible for controlling the liquid level. It may be possible to manipulate the inlet flow rate by installing a control valve in the inlet line, installing an overflow weir, or adding a make-up stream to the system. The process unit downstream of the holding tank may require a set flow to operate, and in this case, the outlet flow rate cannot be a manipulated variable, so an alternative strategy needs to be devised. Temperature, pressure, and concentration are among the other variables that are frequently controlled in chemical processes. Figure 3.12 shows a control scheme for maintaining the temperature of a reactor at the desired setpoint by manipulating the flow rate of the heat transfer medium in the reactor jacket.

In all cases, the approach for process control is the same: develop the transfer function to quantify the relationships, and design an algorithm to change the manipulated variable to counteract the disturbance. Controlling batch processes involves additional complexity of time dependency. Here the *setpoint* (the desired value of the controlled variable) may

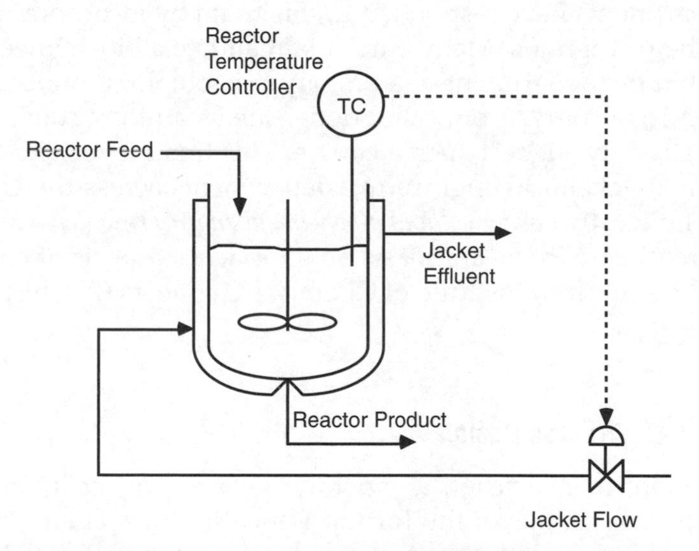

Figure 3.12 Temperature control for a reactor.
Source: Bequette, B. W., *Process Control: Modeling, Design and Simulation*, Prentice Hall, Upper Saddle River, New Jersey, 2003.

change with time, unlike the continuous processes. Despite this complication, a control algorithm and the control strategy can be developed along the same lines.

3.3.1.5 Elective Courses

Apart from the essential courses, a chemical engineering student will typically have anywhere from three to five technical courses in chemical engineering and related areas as electives. The course offerings vary among institutions depending on the faculty expertise and interests. Materials related courses, such as courses in corrosion, polymer engineering, and ceramic materials, are common if the chemical engineering program is closely aligned with material science and engineering. Several chemical engineering departments have expertise in bio-related areas, and biochemical engineering, bioenergy, and biomedical courses are the available electives. Students may also take as electives relevant courses in environmental engineering, petroleum engineering, semiconductors, and nuclear engineering.

Engineering activities related to the biological or life-sciences area have grown exponentially in recent times, as was seen from the graph of engineering graduates in Chapter 1. A large number of chemical engineering departments have responded to this trend by incorporating relevant courses in their programs. Many curricula mandate a bioengineering course offered within the department or a biology, microbiology, molecular biology, or biochemistry course from other departments and programs, while several others strongly suggest such a course. This trend is reflected both in the names of many chemical engineering departments across the United States, which are now called *chemical and biochemical engineering departments* (or a variation thereof), and in the emphasis on life-science aspects of chemical engineering in the American Institute of Chemical Engineers (AIChE) definition stated in Chapter 1.

3.3.1.6 Process Design

As mentioned earlier, a capstone design project in the final year of the program completes the formal education of a chemical engineer. This is accomplished at the University of Idaho through a two-semester sequence of courses called *Process Analysis and Design I* and *II*. Typically, a student is assigned a comprehensive project that requires him/her to apply the concepts learned in other courses and present a detailed solution. The student invariably is required to submit a technical document and make a presentation to the class, college, or a wider audience. The project may be an individual or group assignment and may involve building working prototypes and conducting demonstration experiments. The scope of the project may be as wide as designing a complete plant for the production of a chemical, or it may be very specific and focused, such as removing a particular contaminant from a specific effluent stream.

Students typically learn the fundamentals of process design and plant economics in the initial part of the course at the beginning of the final year. Cost estimation of process units, accounting for capital and operating costs, cash flow, and return on investment analysis are some to the topics covered in the course [11]. Students learn the techniques to assign economic value to the engineering activities. Students may also get experience working with commercial process simulation software used in industry. The design course is thus the final step in preparing a student for a chemical engineering career. It is not unusual for a private industry or a public sector employer of engineers to agree to provide actual engineering problems related to their businesses as senior design problems for student groups. Such entities may participate in the design process by providing funding or advice as appropriate. Such real-world problems are often especially beneficial to the learning process.

Students enrolling in these advanced courses should have the necessary academic preparation, which is accomplished through the fundamental chemical engineering and engineering science courses.

3.3.2 Fundamental Chemical Engineering Courses

The foundation for the advanced chemical engineering courses is laid in the second and third (sophomore and junior) years of the program through two fundamental chemical engineering courses: *Material and Energy Balance* and *Chemical Engineering Thermodynamics*. Both these courses are described in the following sections.

3.3.2.1 Material and Energy Balance

Scheduled in the second year of the program, this course is possibly the most important course in the chemical engineering curriculum. The title of course is generally Material and Energy Balance, but it may sometimes be called *Chemical Engineering Principles*, *Introduction to Chemical Processes*, or a variation thereof. Simply put, the course is aimed at teaching the students the techniques to conduct the material and energy audits on the process. Material balance involves accounting for quantities of materials flowing in and out of a process unit, a sequence of units, and the entire plant. Energy balance is, similarly, the quantification of energy flows in and out of the system for which this analysis is conducted. In essence, the course teaches students how to apply the principles of *conservation of mass* and *conservation of energy* to units in chemical processes. These units may involve reacting or nonreacting systems, as well as single-phase and multiphase systems. Figure 3.13 presents an overview of this course.

Typically, the course starts with the concept of material balances and application of this concept to a single step in a simple, nonreacting system. Subsequent topics include complex combinations of nonreacting units, a single unit involving reaction(s), a combination of reacting and nonreacting units, energy effects not involving reaction, and energy effects in reacting systems [12].

A student, after successful completion of the course, should be able to determine accurately the mass, volume, temperature, and composition of each material stream flowing into and out of a process unit. The process occurring within this unit may be simple mixing, a physical separation, or a

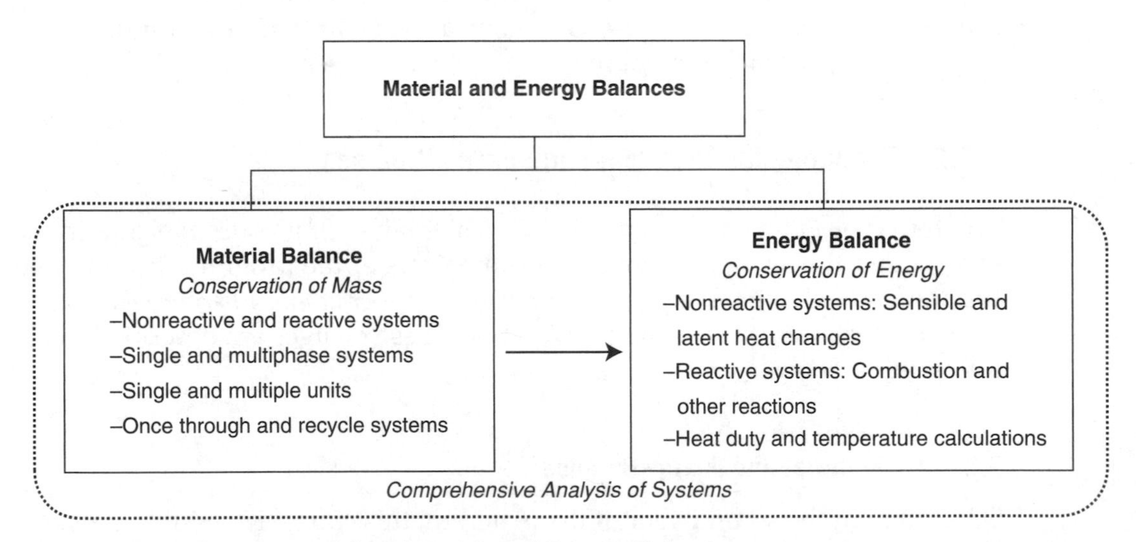

Figure 3.13 Overview of the Material and Energy Balance course.

complex reaction. The process streams may be solid, liquid, gaseous, or any combination thereof. Similarly, the student should be able to determine accurately the energy flows into and out of the units.

Quantifying the material and energy streams connected to a process unit is a prerequisite to designing that unit. These design principles and techniques are taught in the advanced chemical engineering courses previously described. It is clear that unless the student acquires competence in tackling the material and energy balances, he/she will not be in a position to learn the advanced topics in chemical engineering. The Material and Energy Balance course provides an indicator of the ability of a student to be a chemical engineer, often serving as a "gate" for continuation in the chemical engineering program.

3.3.2.2 Chemical Engineering Thermodynamics

A course in engineering thermodynamics is required of students in all engineering disciplines. In addition, chemical engineering students take, generally in the first semester of the third (junior) year, a specialized thermodynamics course called Chemical Engineering Thermodynamics. Thermodynamics is the engineering science that deals with the interconversion between work and energy. The concept of equilibrium is integral to thermodynamics, and the essence of Chemical Engineering Thermodynamics is the study of equilibrium phenomena in chemical systems. Equilibrium can be understood to be the state of a system from which no change is expected; that is, the system is in a stable state. It follows that any system not at equilibrium will have the tendency to move toward equilibrium. The driving force for any process is the departure from equilibrium.

Chemical Engineering Thermodynamics helps students define equilibrium in terms of thermodynamic quantities, such as enthalpy, entropy, free energy, and so on, and relate these thermodynamic quantities to measurable system properties, such as pressure, temperature, volume, and composition [13]. Students learn about the changes in thermodynamic quantities associated with various chemical processes and how to quantify these changes from the volumetric behavior of substances. Volumetric behavior refers to the pressure-volume-temperature relationship for the substance. The ideal gas law is the simplest equation describing the volumetric behavior. Most substances, as they are not ideal gases, require more complex equations, and students learn about nonideal behavior and equations that describe this behavior.

Students learn to apply the laws of thermodynamics to chemical systems—pure substances and mixtures, single phase and multiphase, reacting and nonreacting. Figure 3.14 provides an overview of the course.

Concepts learned in Chemical Engineering Thermodynamics are applied in the advanced courses, particularly separation processes and kinetics, and provide the theoretical foundation for process and equipment design.

3.3.3 Engineering Science Courses

The second (sophomore) year of the curriculum also features the engineering science courses that are common to nearly all engineering disciplines. Principles and concepts learned in these courses serve as prerequisites for higher-level engineering courses. A chemical engineering student also gains an exposure to other engineering disciplines and is in a position to interface with civil, electrical, and mechanical engineers as he/she would invariably be required to do throughout his/her professional career.

3.3.3.1 Fluid Mechanics

Piping is the most ubiquitous feature of a chemical plant. Chemical processes typically involve large material flows, and it is essential for a student to understand the physical phenomena occurring in fluid systems. The *Fluid Mechanics* course teaches the students the application of

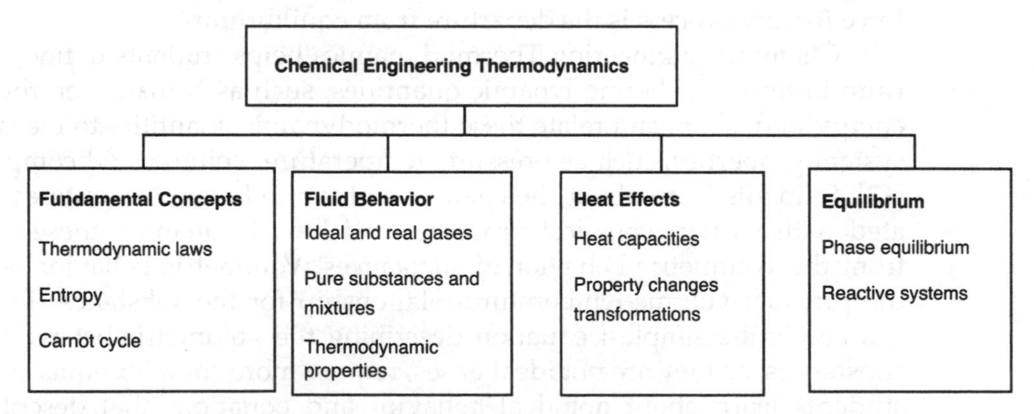

Figure 3.14 Overview of the Chemical Engineering Thermodynamics course.

conservation of mass and energy principles to fluid systems. The two components of Fluid Mechanics are *fluid statics*, the phenomena associated with fluids at rest, and *fluid dynamics*, the phenomena associated with fluids in motion. Fluid systems are analyzed at a macroscopic level; that is, in terms of observable bulk properties of the material. The energy balances generally focus on the mechanical energy—kinetic and potential—of the system, with thermal energy contributions playing an insignificant role.

Students gain an understanding of forces acting on fluid elements in stationary and moving fluids. This understanding is useful in determining the forces and pressure in stationary fluids as well as energy and power requirements in flow systems. Substantial costs are incurred in chemical processes in simply moving material from one process unit to another. Figure 3.15 provides an overview of the course.

The concepts learned in Fluid Mechanics are prerequisite for the transport phenomena courses, which analyze the systems from a microscopic or molecular level. The Fluid Mechanics course helps the students understand the macroscopic behavior of the fluids. The transport phenomena courses delve into the explanation of this observed behavior through the analysis of microscopic processes occurring in the fluid.

3.3.3.2 Engineering Thermodynamics

As mentioned previously, *thermodynamics* is the branch of science that deals with the interconversion of energy and work. Engineers, at a very basic level, are individuals who deal with engines—machinery and processes—to

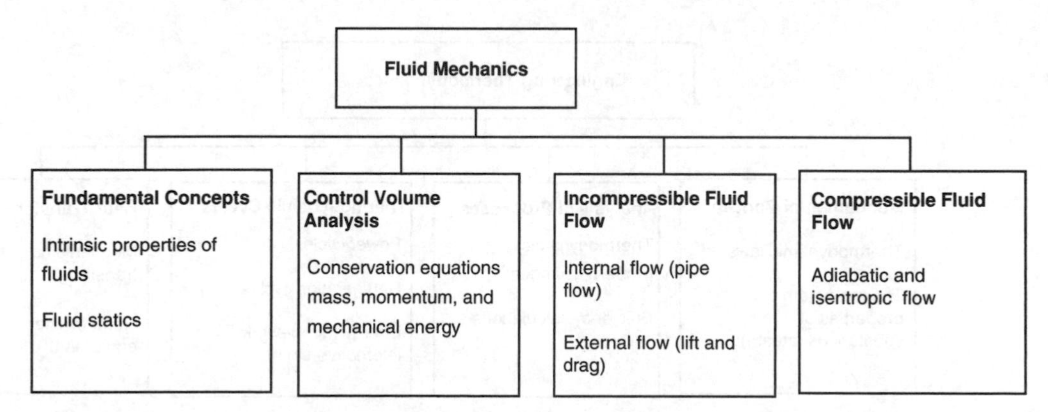

Figure 3.15 Overview of the Fluid Mechanics course.

obtain useful work for the benefit of the society. This work is obtained at the expense of energy, and engineers need to have fundamental understanding of the concepts that govern the relationship between energy and work. *Engineering Thermodynamics* teaches students the *laws of thermodynamics* and applications of these laws to various systems. Students also learn about the thermodynamic quantities (enthalpy, entropy, free energy, etc.), as mentioned earlier, and the changes in these quantities in various types of processes. The process may be *isothermal* (occurring at constant temperature), *isobaric* (at constant pressure), or *isochoric* (at constant volume). Engineers construct power conversion cycles consisting of combinations of these and other processes. These concepts are then applied to the analysis of energy or power conversion cycles that involve a cyclic sequence of different processes. The course also covers macroscopic analysis of heat transport processes. An overview of the Engineering Thermodynamics course is shown in Figure 3.16.

The concepts covered in Engineering Thermodynamics are prerequisites for courses such as Chemical Engineering Thermodynamics and Transport Phenomena. It can be seen that the topics covered in the Engineering Thermodynamics course are more general in nature than those in the Chemical Engineering Thermodynamics course. Although some chemical engineering programs conform to this arrangement of the two courses, several other programs may simply have two chemical engineering thermodynamics courses, with the topical areas corresponding roughly to those described previously. If the first engineering thermodynamics course is taught by the faculty in the chemical engineering program with the enrollment restricted primarily to the chemical engineering

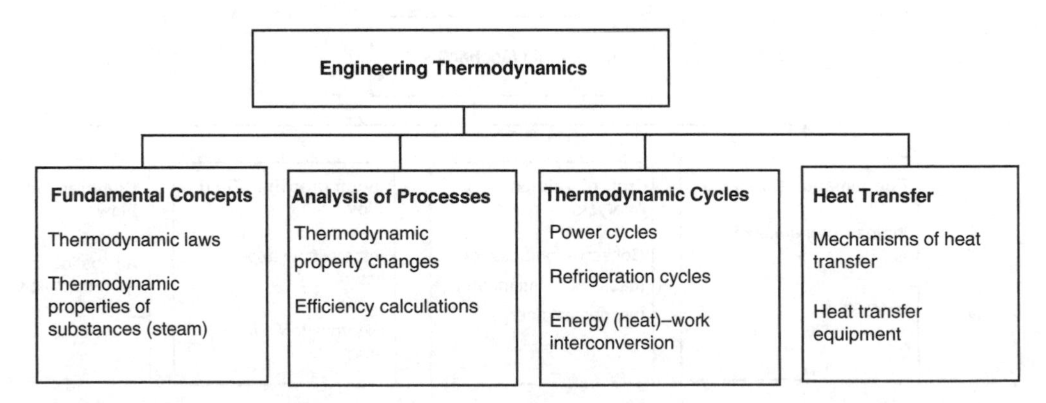

Figure 3.16 Overview of the Engineering Thermodynamics course.

majors, then typically it is named *Chemical Engineering Thermodynamics I*. If the course is offered by any faculty member in the college of engineering, with enrollment open to any engineering student, it typically appears as *Engineering Thermodynamics*. Regardless of the specific arrangement, the two courses function to provide discussion of fundamental concepts of engineering thermodynamics and their application to chemical engineering systems.

3.3.3.3 Engineering Statics and Electrical Circuits

Chemical engineering students also typically are exposed to the concepts related to civil engineering and electrical engineering through courses such as *Engineering Statics* and *Electrical Circuits*. The students learn about force balance on rigid bodies, equilibrium, strength of materials and structures, and so on, in the Engineering Statics course. Students learn how to analyze trusses and frames and other load-bearing structures. This knowledge is useful in the design of chemical process equipment, where the chemical engineer must take into consideration the load demand an equipment will place on the supporting structure.

Similarly, the Electrical Circuits course exposes the chemical engineering students to concepts related to electrical systems. Students learn to analyze the steady-state and transient behavior of electrical energy and power systems. This knowledge is also useful in the design of processes and equipment where the chemical engineer must take into account electrical requirements imposed by the unit.

3.3.3.4 Computer Programming

Most chemical engineering programs require students to take a course in computer science, typically a course in a programming language. Historically, most engineers studied FORTRAN programming, but the emphasis has shifted to other newer languages and software in recent years. Students generally take a course in programming languages such as Visual Basic, C/C++, MATLAB, or other software languages. Irrespective of the specific language, students learn about algorithm development and code formulation in the course. These skills are invaluable for any engineer, as growing computational power enables an engineer to tackle problems of increasing complexity.

Success in these fundamental courses, in turn, is based on basic science and mathematics courses, scheduled in the first two years of the curriculum. These courses are described next.

3.3.4 Fundamental Science and Mathematics Courses

Chemical engineering students take several chemistry, physics, and mathematics courses during the first (freshman) and second (sophomore) years.

3.3.4.1 Chemistry Courses

Most engineering students in other disciplines take perhaps one course in chemistry, typically in the first semester of the first (freshman) year. Chemical engineering students take a significantly larger number of chemistry courses. The courses in the first year of study expose students to concepts in general and inorganic chemistry. The next two to three semesters involve instruction in organic chemistry and physical chemistry. After completion of these courses a student should have knowledge of the following areas:

- Kinetics and thermodynamics
- Equilibrium
- Acids and bases
- Electrochemistry
- Nuclear chemistry
- Synthesis and properties of organic compounds
- Biochemistry
- Quantum mechanics

Each course consists of lecture and laboratory components, giving students an opportunity to gain both conceptual and practical knowledge.

In addition to these required courses, students may take more chemistry courses to satisfy the science elective requirements. Chemical engineering students in many institutions may be able to work these additional chemistry courses into their study plans to satisfy requirements for the bachelor's degree in chemistry.

3.3.4.2 Physics Courses

Chemical engineering students typically take two physics courses, generally completed by the third semester. The topics covered in these courses generally include the following:

- Kinematics and dynamics
- Newton's laws
- Friction
- Static equilibrium
- Work and energy
- Gravity and central forces
- Momentum
- Electrical fields and potentials
- Magnetism
- AC/DC circuits, capacitance, and inductance

As with chemistry courses, these courses also have lecture and laboratory components.

3.3.4.3 Mathematics Courses

Chemical engineering students, similar to other engineering students, typically take a sequence of three semester courses in mathematics covering various topics in analytical geometry and calculus. The topics covered in these courses include functions, limits and continuity, series, integration and differentiation, vectors, algebraic and transcendental equations, numerical techniques, conics and solid geometry, and so on. These courses are generally prerequisites for a course on differential equations, which is one of the most important courses in the education of an engineer. Students learn about first- and higher-order differential equations; initial and boundary value problems; solution techniques for equations, including series solutions and Laplace transforms; and systems of linear equations. These differential equations are representative of the mathematical models of systems encountered in the engineering field.

The relationship between various science and engineering courses that are essential for the education of a chemical engineer is summarized in Figure 3.17.

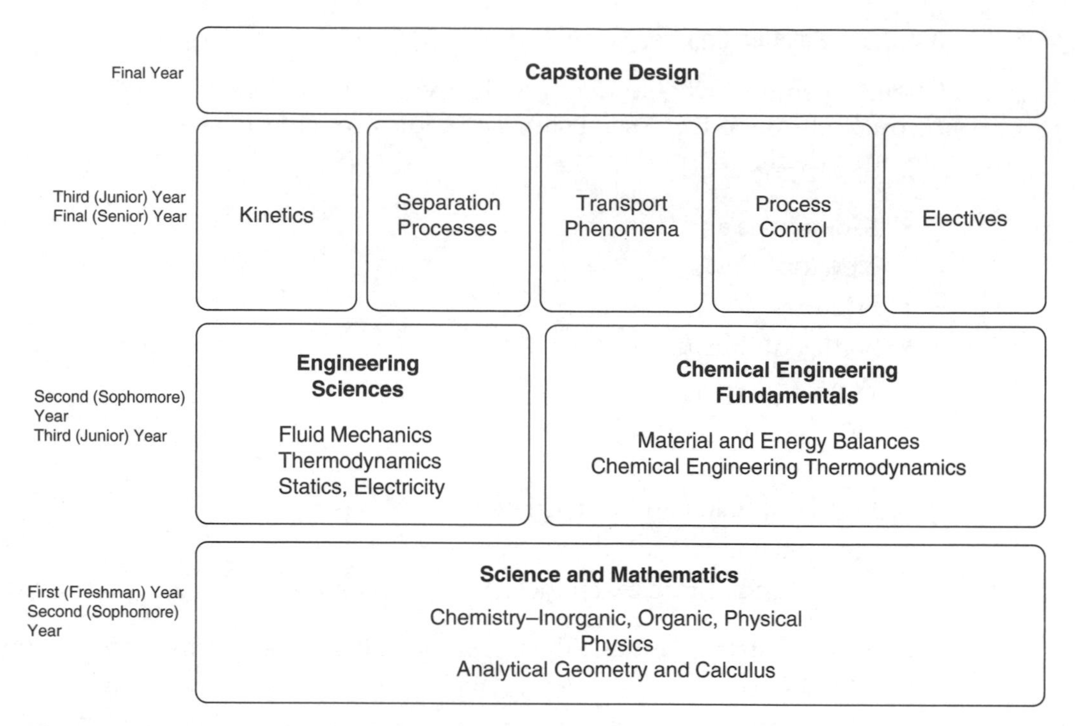

Figure 3.17 Hierarchy of technical courses in chemical engineering curriculum.

3.3.5 General Education Courses

In addition to these technical courses, a proper number of general education, humanities, and social science courses is absolutely essential to complete the education of a chemical engineer. Two of these courses play almost as important a role in an individual's career as any of the technical courses: *economics* and *communication*.

A course in economics is essential for the engineer to understand the ultimate driving force for most, if not all, professional activities he/she will be involved in. Economic considerations play a major role in evaluation of alternative technologies and processes and often dictate the ultimate choice. The course provides an engineer with not only an understanding of the interplay of economics and technology but also a perspective about the role of a chemical engineer in a societal context.

Competence in reaction engineering, thermodynamics, transport phenomena, and separations is viewed as an essential qualification by the industry. However, an ability to communicate is an equally important skill that an employer looks for in a prospective employee [14]. Chemical engineers need to interact and communicate with a broad range of people having varying levels of technical knowledge. They need to communicate with chemical and other engineers, technical managers and administrators, sales/marketing and finance personnel, and plant operators and technicians, among others. The chemical engineer should be able to convey his/her ideas clearly and succinctly to all these people. A chemical engineer will quickly discover that technically sound, logical ideas that benefit the process need to be articulated properly even in a peer group of chemical engineers for them to accept it. A course that helps develop written and oral communication skills is a must for a budding chemical engineer.

Economics and communication are probably the two most important general education courses, but the value of the other general education courses cannot be underestimated. The humanities and social science courses in disciplines such as philosophy, history, sociology, and so on are absolutely essential for the holistic education of an individual. These courses provide a moral and ethical framework for one's thoughts and actions, present the context for the development of societies, and promote an understanding and appreciation of the diversity of viewpoints, and they provide many other such benefits that can only lead to a better chemical engineer.

3.4 Summary

The education of a chemical engineer starts with the fundamental chemistry, physics, and mathematics courses. These courses lead the student into the engineering science and fundamental chemical engineering courses. The advanced chemical engineering courses teach students how to design the chemical process equipment and control the process to build on these foundation courses. The capstone design course enables the student to synthesize the technical knowledge to solve a problem of value to society. A student completing these studies will have experienced an intellectually challenging curriculum, one that will lead to a professionally satisfying and rewarding career and one that prepares him/her to be a contributing and constructive member of society.

References

1. Leigh, G. H., *The World's Greatest Fix: A History of Nitrogen and Agriculture*, Oxford University Press, Oxford, England, U.K., 2004.
2. Smil, V., *Enriching the Earth: Fritz Haber, Carl Bosch and the Transformation of World Food Production*, MIT Press, Cambridge, Massachusetts, 2001.
3. Ertl, G., "The Arduous Way to the Haber-Bosch Process," *ZAAC*, 2012, pp. 487–489.
4. Doraiswamy, L. K., and D. Üner, *Chemical Reaction Engineering: Beyond the Fundamentals*, CRC Press, Boca Raton, Florida, 2014.
5. Austin, G. T., *Shreve's Chemical Process Industries*, Fifth Edition, McGraw-Hill, New York, 1984.
6. Fogler, H. S., *Elements of Chemical Reaction Engineering*, Fourth Edition, Prentice Hall, Upper Saddle River, New Jersey, 2005.
7. Wankat, P. C., *Separation Process Engineering*, Third Edition, Prentice Hall, Upper Saddle River, New Jersey, 2012.
8. Luyben, W. L., and L. A. Wenzel, *Chemical Process Analysis: Mass and Energy Balances*, Prentice Hall, Englewood Cliffs, New Jersey, 1988.
9. Thomson, W. J., *Introduction to Transport Phenomena*, Prentice Hall, Upper Saddle River, New Jersey, 2000.
10. Bequette, B. W., *Process Control: Modeling, Design and Simulation*, Prentice Hall, Upper Saddle River, New Jersey, 2003.
11. Peters, M. S., K. D. Timmerhaus, and R. E. West, *Plant Design and Economics for Chemical Engineers*, Fifth Edition, McGraw-Hill, New York, 2011.
12. Himmelblau, D. M., and J. B. Riggs, *Basic Principles and Calculations in Chemical Engineering*, Eighth Edition, Prentice Hall, Upper Saddle River, New Jersey, 2012.
13. Kyle, B. G., *Chemical and Process Thermodynamics*, Third Edition, Prentice Hall, Upper Saddle River, New Jersey, 1999.
14. CEP Update, "How Well Are We Preparing ChE Students for Industry?" *Chemical Engineering Progress*, Vol. 110, No. 4, 2014, pp. 4–6.

Problems

3.1 Compare the bachelor's degree curriculum at your institution with the curricula from two other schools. What are the similarities? Where are the major differences?

3.2 What are the different electives available at your institution? What subject areas would you like to see offered as electives? Why?

3.3 Compare microbial and synthetic nitrogen fixation processes.

3.4 Based on the description of batch and continuous reactors, explain the possible operation of a semibatch reactor.

3.5 How does a stripping operation compare with an absorption operation? What are the similarities and differences?

3.6 Explain what is meant by controlled and manipulated variables, with the help of an example from everyday life, such as the operation of cruise control on a car.

3.7 How does a course in engineering thermodynamics prepare students for the course in chemical engineering thermodynamics?

3.8 Describe the concept of equilibrium with the help of an example.

3.9 The human body is an amazing machine that continually carries out several separation operations. Give an example of a membrane separation occurring in the human body.

3.10 Discuss the significance of the contribution of the courses in humanities and social sciences in engineering education.

CHAPTER 4

Introduction to Computations in Chemical Engineering

It is unworthy of excellent men to lose hours like slaves in the labor of calculation which could be safely relegated to anyone else if machines were used.

—Gottfried Wilhelm von Leibniz[1]

Chemical engineering, like all engineering disciplines, is a quantitative field; that is, it requires accurate solutions of problems having high mathematical complexity. A chemical engineer must be able to *model*—develop quantitative mathematical expressions that describe the processes and phenomena—any system of interest, and *simulate*—solve the equations—the model. The solutions so obtained allow the engineer to design, operate, and control the processes. The courses described in Chapter 3, "Making of a Chemical Engineer," provide the students with the theoretical basis for modeling the processes. The nature of the resulting equations and tools used for solving the equations are presented in this chapter.

4.1 Nature of Chemical Engineering Computational Problems

Chemical engineers deal with a multitude of equations ranging in complexity from simple linear equations to highly involved partial differential equations. The solution techniques accordingly range from simple calculations to very large computer programs. The classification of the problems based on the mathematical nature is presented in the following sections.

1. Leibniz's fame as the co-inventor of calculus overshadows his contributions in many other fields, including those in the field of computations. He was one of the earliest pioneers in the development of mechanical calculators. Quotation source: http://www-history.mcs.st-and.ac.uk/Quotations/Leibniz.html

4.1.1 Algebraic Equations

Algebraic equations comprise the most common group of problems in chemical engineering. *Linear algebraic equations* are algebraic equations in which all the terms are either a constant or a first-order variable [1]. The straight line is represented by a linear algebraic equation. Linear algebraic equations are often encountered in phase equilibrium problems associated with separation processes. Figure 4.1 is a representation of one such separation operation, wherein a high-pressure liquid stream is fed to a flash drum where the system pressure is reduced, resulting in the formation of a vapor and a liquid stream that exit the drum. The compositions of the liquid and the vapor stream depend on the process conditions, and a chemical engineer has to calculate these compositions.

The governing equations for the system follow:

$$\sum_{i=1}^{n} x_i = 1 \tag{4.1}$$

$$\sum_{i=1}^{n} y_i = 1 \tag{4.2}$$

$$y_i = K_i x_i \text{ for } i = 1..n \tag{4.3}$$

Equations 4.1 and 4.2 state that the mole fractions of all components, numbering n, in each phase add up to 1. x_i and y_i represent the mole fractions

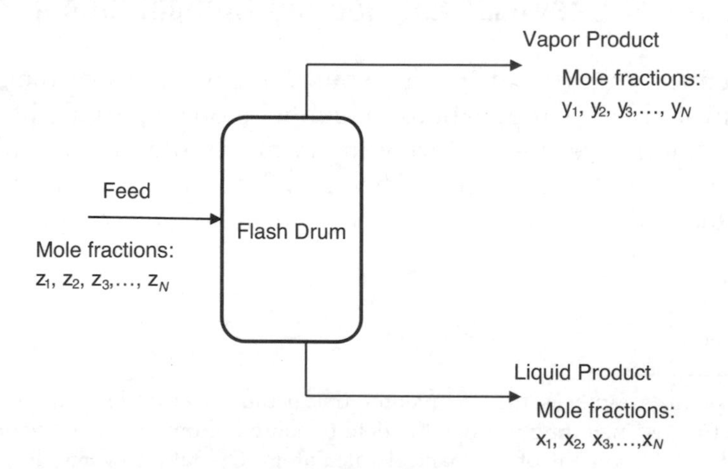

Figure 4.1 Operation of a flash drum.

of component i in the outlet liquid and gas phases, respectively. The mole fractions in the feed stream are denoted by z_i. (Typically, x is used to represent the mole fraction when the phase is liquid, and y is used when the phase is gaseous.) These two equations should be intuitively clear, as the mathematical statements of the concept that all fractions of any quantity must add up to the whole. Equation 4.3 is actually a system of n equations relating the mole fraction of a component in the gas phase to the mole fraction of the same component in the liquid phase. K_i is a characteristic constant for component i and is dependent on pressure, temperature, and the nature of the component mixture. Solution of this system of equations allows us to calculate the compositions of the two different phases, which is necessary for designing the separation scheme for the mixture. Each term in the system of equations is linear (variables having power of 1) in x or y.

A similar system of equations is used to model a stagewise gas-liquid contactor, such as a distillation column, described in Chapter 3. Figure 4.2 represents a distillation column containing N equilibrium stages [2]; the vapor and liquid inlet and outlet flows can be seen in the figure for stage k.

The material balances for each component yield the following system of n equations for stage k:

$$V_{k-1}y_{i,k-1} + L_{k+1}x_{i,k+1} = V_k y_{i,k} + L_k x_{i,k} \text{ for } i = 1..n \tag{4.4}$$

V and L represent the molar flow rates of the vapor and liquid stream, respectively. The subscripts for these flow rates represent the stage *from* which these flows exit. For example, V_k and L_k are the vapor and liquid flow rates exiting stage k, respectively. L_{k+1} is the liquid flow rate exiting stage $k + 1$ and entering stage k, and V_{k-1} is the vapor flow rate exiting stage $k - 1$ and entering stage k. The mole fractions are doubly subscripted variables, the first subscript representing the component, the second one the stage. Equation 4.4 is the mathematical representation of the steady-state nature of the system for each component: the amount of component i entering the stage through vapor and liquid flows is the same as the amount leaving through the exiting vapor and liquid flows. Each stage is assumed to be an equilibrium stage; that is, the exiting vapor and liquid flows are in equilibrium with each other. This allows us to utilize the equilibrium relationships of the form shown by equation 4.3 to complete the system description. The total number of equations for the entire column is $N \times n$,

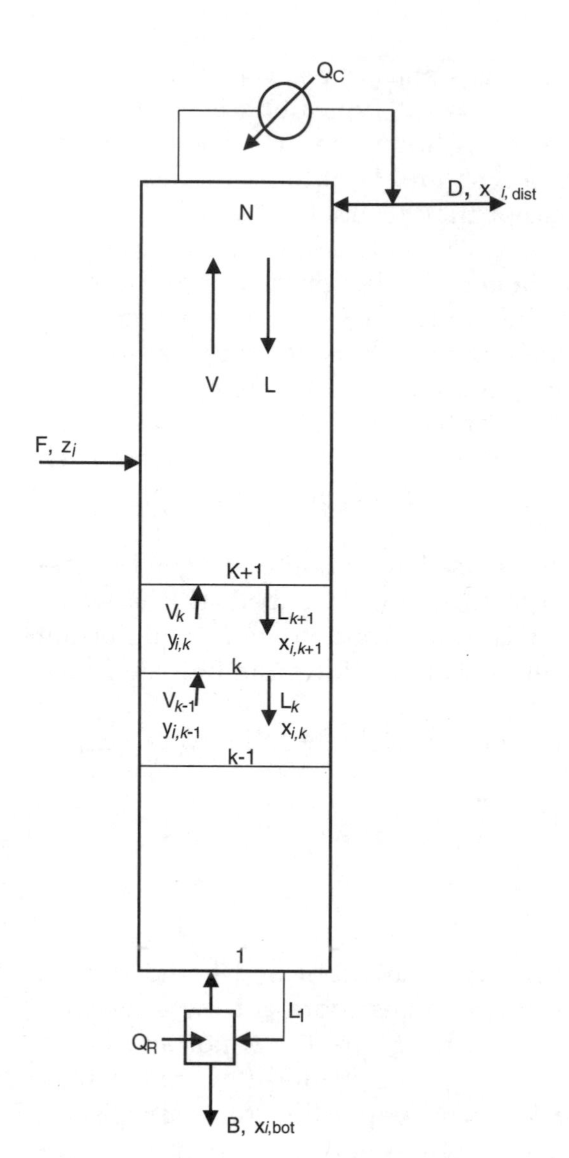

Figure 4.2 Distillation column—stagewise operation.

Source: Adapted from Wankat, P. C., *Separation Process Principles*, Third Edition, Prentice Hall, Upper Saddle River, New Jersey, 2012.

which can be significantly large depending on the number of components present in the process stream and the number of stages needed to obtain the desired separation.

Algebraic equations encountered in chemical engineering can also be *polynomial equations*; that is, they can have variable orders greater than one. Equation 4.5 represents a typical polynomial equation of interest to chemical engineers:

$$aV^3 + bV^2 + cV + d = 0 \tag{4.5}$$

This equation is an example of a *cubic equation of state*, V being the volume of the substance under the given conditions of temperature (T) and pressure (P). Constants a, b, c, and d are functions of the system pressure, temperature, number of moles, and fluid properties. An equation of state represents the relationship between the system temperature, pressure, and volume; the ideal gas law represented by the mathematical expression $PV = nRT$ is the simplest of the equations of state. These equations of state are further used in thermodynamic calculations involving interconversion between energy and work, and phase equilibrium. It is readily apparent that an accurate equation of state is critical for superior process design and performance. Unfortunately, the volumetric behavior of most substances does not conform to the ideal gas law, and more complex equations are needed for accurately describing the P-V-T relationships for these substances. The cubic equations of state represent one of the developments addressing this need for improved accuracy. Equation 4.6 is an example of the cubic equation of state and is called the *van der Waals equation* [3].

$$\left(P + \frac{an^2}{V^2}\right) \cdot (V - nb) = nRT \tag{4.6}$$

In this equation, a and b are constants characteristic of the substance, and n is the number of moles present in the system.

Several other more complex equations have also been developed, many of them polynomial in nature. A chemical engineering student encounters polynomial equations in practically every subject described in Chapter 3.

4.1.2 Transcendental Equations

Many of the equations in chemical engineering involve functions of variables more complex than simple powers. An equation containing exponential, logarithmic, trigonometric, and other similar functions is not amenable to solution by algebraic means—that is, by simple addition, multiplication, or root extraction operations. Such equations "transcend" algebra and are called *transcendental equations* [4]. Equation 4.7, the *Nikuradse equation*, often used in fluid flow calculations, is an example of a transcendental equation [5].

$$\frac{1}{\sqrt{f}} = 4.0\log\left\{\mathrm{Re}\sqrt{f}\right\} - 0.40 \tag{4.7}$$

Re in the equation represents *Reynolds number*, a dimensionless quantity of enormous significance in fluid mechanics and transport phenomena. The Nikuradse equation allows us to calculate f, the *friction factor*, a quantity that further leads to the estimation of pressure drop for a flowing fluid and, ultimately, the power requirements for material transfer.

Equation 4.8 is another example of a transcendental equation that is used in the design of chemical reactors [6].

$$X_A = \frac{\tau A e^{-E/RT}}{1 + \tau A e^{-E/RT}} \tag{4.8}$$

X_A represents the conversion (extent of reaction) of the reactant A, τ the residence time (the time spent by the fluid in the reactor), and A and E the characteristic parameters that describe the rate of reaction. The equation can be used to calculate one of the three quantities X_A, τ, or T when the other two are specified.

Many processes involve consecutive chemical reactions that can be represented by the equation $A \rightarrow R \rightarrow S$. A is the starting reactant, which upon undergoing the reaction yields the specie R, which is often the desired product. However, R may undergo further reaction forming S. The typical concentration profiles for the three species in a reactor as a function of time are shown in Figure 4.3. As can be seen, the concentration of A decreases continuously, while that of S increases continuously. The concentration of the desired product R increases first, reaches a maximum, and then starts decreasing. The concentration-time relationship for R when both the reactions are first order[2] with respect to the reactants is shown in equation 4.9 [7].

$$C_R = C_{A0} \cdot \frac{k_1}{k_2 - k_1}\left[e^{-k_1 t} - e^{-k_2 t}\right] \tag{4.9}$$

2. A first-order reaction is one in which the rate of reaction is proportional to the concentration of the reactant. These concepts are presented in more detail in Chapter 9, "Computations in Chemical Engineering Kinetics."

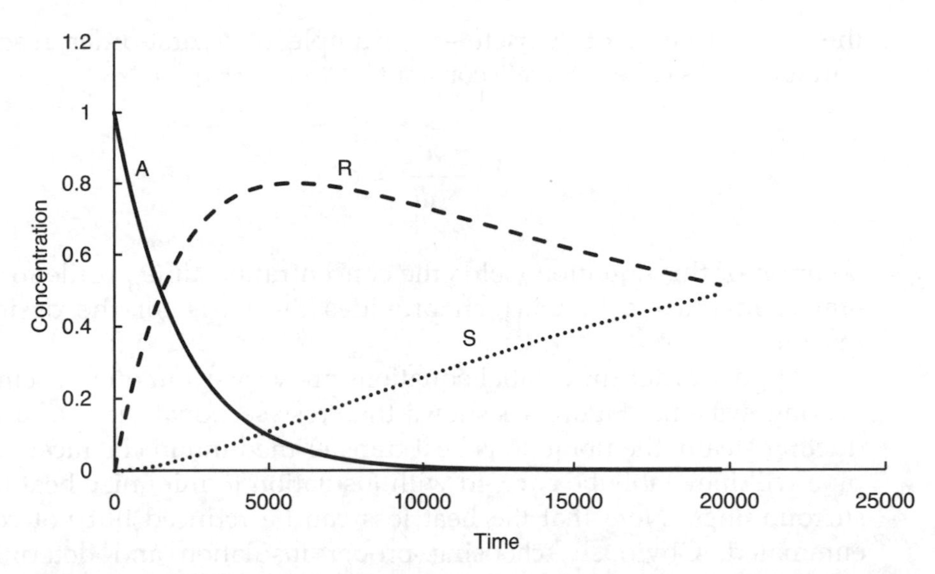

Figure 4.3 Concentration profiles of species for the consecutive reaction scheme A → R → S.

Here, C_R is the concentration of R, C_{A0} is the initial concentration of A, and k_1 and k_2 are the rate constants for the two reactions. Calculating the concentration of R at any specified time, when the rate constants and initial concentration of A are known, is straightforward. However, calculation of time needed to achieve a certain specified concentration of R is more challenging and requires use of techniques needed for the solution of transcendental equations.

4.1.3 Ordinary Differential Equations

Modeling—developing a set of governing equations—of systems of interest to chemical engineers often starts with defining a *differential element* of the system. This differential element is a subset of the larger system, but with infinitesimally small dimensions. All the processes and phenomena occurring in the larger system are represented in the differential element. The modeling approach involves writing conservation of mass and/or conservation of energy equations for the differential element. These equations yield ordinary differential equations when all the quantities are functions of a single independent variable. For example, equation 4.10 is a first-order differential equation relating the rate of change of concentration to time in a chemical reaction [6]. The equation indicates that the rate at which the concentration of species A, C_A, changes with time t is linearly dependent on the concentration of A itself—an example of a first-order reaction. The parameter k is called the rate constant.

$$-\frac{dC_A}{dt} = kC_A \tag{4.10}$$

Solution of this equation yields the concentration-time profile for the reactant A in the reaction, which provides the basis for the design of the reactor.

Higher-order differential equations are very common in chemical engineering systems. Figure 4.4 shows the cross-sectional view of a pipe conducting steam, the ubiquitous heat transfer medium in chemical plants. The pipe will inevitably be covered with insulation to minimize heat loss to the surroundings. Note that the heat loss can be reduced but not completely eliminated. Obviously, choosing proper insulation and determining the resultant heat loss is extremely important for estimating the energy costs. Heat loss can be calculated from the temperature-distance profiles existing in the system [5].

The governing equation describing the heat transfer for a cylindrical pipe follows:

$$\frac{d}{dr}\left(k_T r \frac{dT}{dr}\right) = 0 \tag{4.11}$$

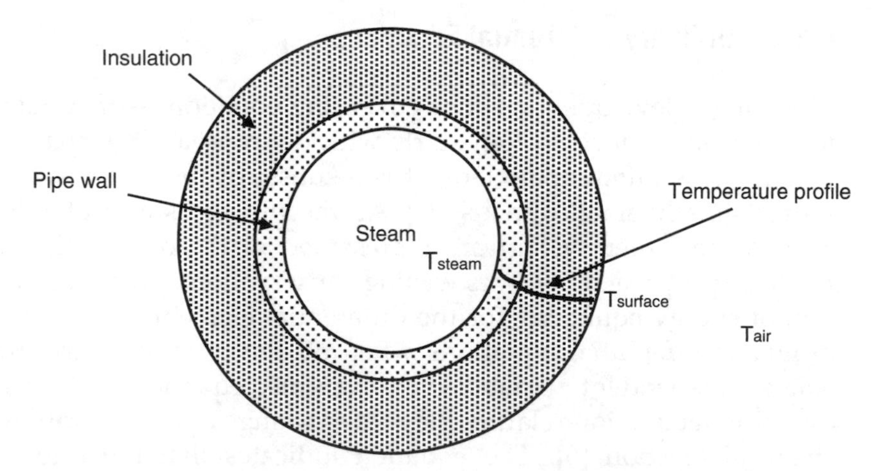

Figure 4.4 Temperature profile for an insulated steam pipe.

Equation 4.11 is a second-order ordinary differential equation that governs the relationship between temperature T and radial distance r from the center of the pipe. k_T is the thermal conductivity of the material, which depends on the temperature. As the temperature varies with respect to the radial position, the thermal conductivity is also a function of the radial position. Solution of this equation yields the temperature profile within that object, which in turn allows us to determine the heat lost to the surroundings.

The solution of differential equations requires specifying values of dependent variable(s) at certain values of the independent variable. These specifications are termed *boundary* conditions (at a specific location, with respect to dimensional coordinate) or *initial* conditions (with respect to time). Complete solution requires as many boundary/initial conditions as the order of the differential equation [4].

Frequently, modeling of a system leads to a set of ordinary differential equations, consisting of two or more dependent variables that are functions of the same independent variable. These equations need to be solved simultaneously to obtain the quantitative description of the system.

4.1.4 Partial Differential Equations

Properties of systems are frequently dependent on, or are functions of, more than one independent variable. Modeling of such systems leads to a partial differential equation [4]. Temperature within a rod, for example, may vary radially as well as axially. Similarly, concentration of a species within a system may depend on the location as well as vary with time. Figure 4.5 shows batch drying of a polymer film cast on a surface. The solvent present in the polymer diffuses through the film to the surface, where it is carried away by an air sweep.

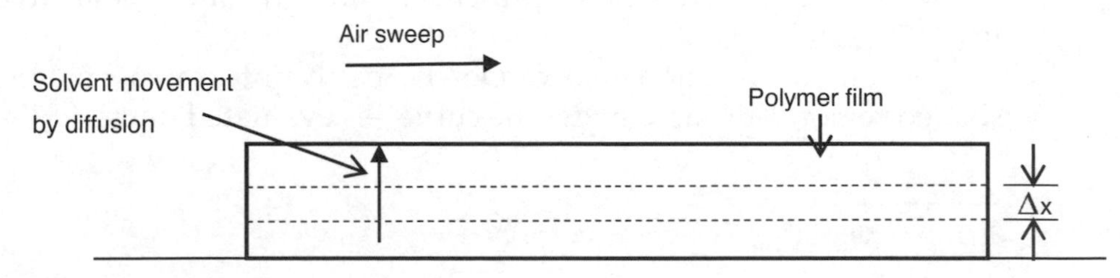

Figure 4.5 Drying of polymer film.

The concentration of the solvent within the film is a function of time as well as distance from the surface. Equation 4.12 is the fundamental equation[3] for governing the solvent mass transport within the film, a partial differential equation that is first order with respect to time t and second order with respect to location x.

$$\frac{\partial C_A}{\partial t} = D_A \frac{\partial^2 C_A}{\partial x^2} \tag{4.12}$$

D_A is the diffusivity of solvent A in the polymer film, which depends on the properties of the system.

The solution of this (and other partial differential equations) requires an appropriate number of specifications (boundary and initial conditions) depending on orders with respect to the independent variables.

4.1.5 Integral Equations

The differential equations representing the behavior of the system are obtained by the application of conservation principles to a differential element. Integration of these differential equations leads to expressions that describe the overall behavior of the entire system. Many of the differential equations can be integrated analytically, yielding algebraic or transcendental equations. However, such analytical integration is not always possible, and numerical computation is necessary for obtaining the integrals [4]. The determination of reactor volume often involves equations of the following form [6]:

$$V = F_{A0} \int_0^{x_{A,final}} \frac{dx_A}{-r_A} \tag{4.13}$$

Here, F_{A0} is the molar flow rate of species A, and $-r_A$ is the rate of reaction, which is a function of conversion X_A. Equation 4.13 is represented by Figure 4.6, where the shaded region represents the integral and is equal to the quantity V/F_{A0}.

When the reaction rate cannot be easily integrated analytically, the shaded region—the area under the curve—is evaluated numerically.

3. This is called the *Fick's second law equation*.

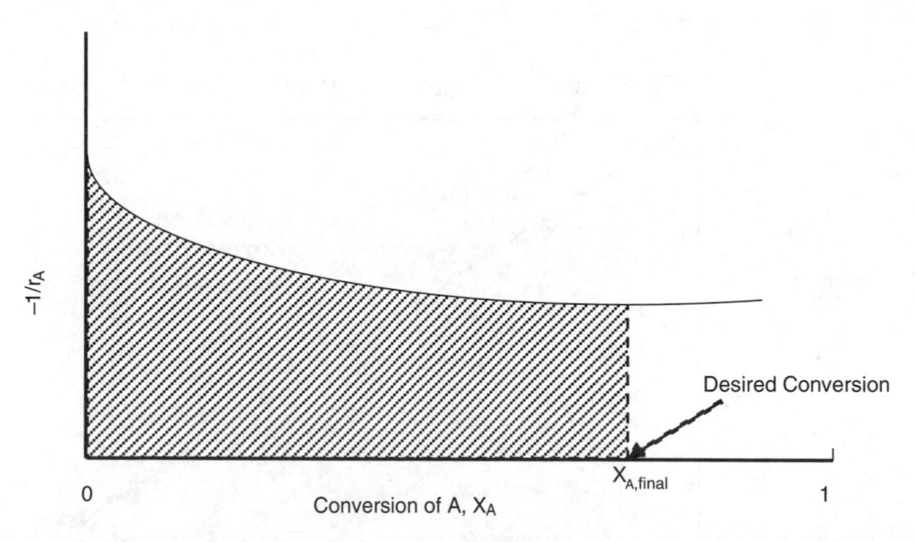

Figure 4.6 Determination of reactor volume.

4.1.6 Regression Analysis and Interpolation

Chemical engineers routinely collect discrete data through various experiments, which they further use for design, control, and optimization. This often requires obtaining the value of the function (or dependent variable) at some value of the independent variable within the domain of experimental data where direct measurement is not available. Regression analysis involves fitting a smooth curve that approximates the data, yielding a continuous function [4]. It is then possible to interpolate—obtain the function value at any intermediate value of the independent variable. It is also possible to extrapolate—obtain the function value at a value of the independent variable that is *outside* the data range used for regression analysis. *Linear regression* involves approximating the data using a straight line, whereas *nonlinear regression* involves using polynomial or transcendental functions for the same purpose. *Multiple regression* involves performing regression analysis involving two or more independent variables that determine the value of the function. For example, equation 4.10 can be integrated to obtain the following mathematical relationship between concentration and time:

$$\ln C_A = \ln C_{A0} - kt \tag{4.14}$$

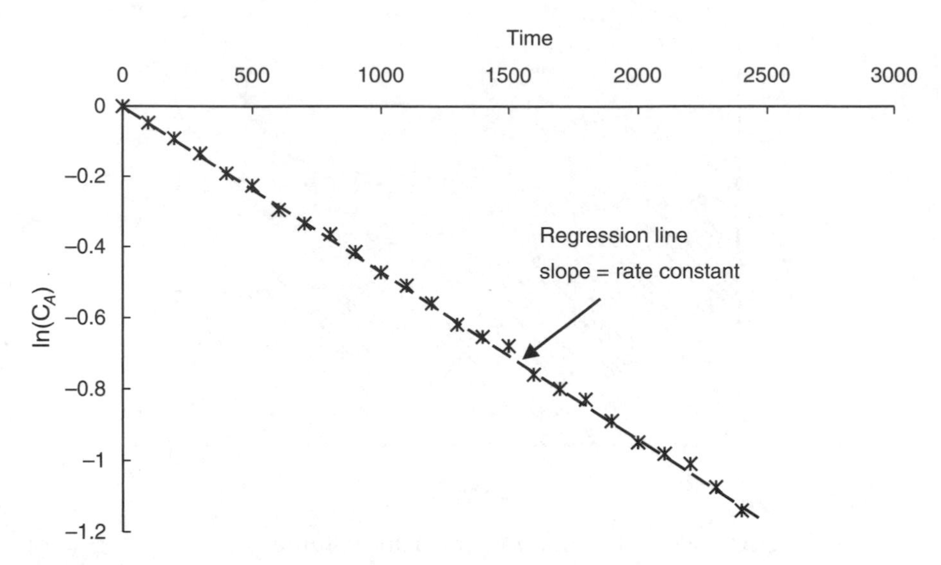

Figure 4.7 Example of linear regression for determination of rate constant.

To determine the rate constant k, experiments are conducted obtaining the concentration-time data and a linear regression carried out between $\ln(C_A)$ and t, as shown in Figure 4.7.

It can readily be seen that a chemical engineer must have skills to deal with and solve problems ranging from simple arithmetic calculations to those requiring highly sophisticated and involved algorithms. Further, the solution must be obtained fairly rapidly for the individuals and organizations to maintain their competitive edge and respond to changing conditions. Section 4.2 presents a brief overview of solution algorithms developed for numerical solutions of different types of problems. Section 4.3 describes the different tools including the machines and software available to chemical engineers to perform these computations.

4.2 Solution Algorithms

The theoretical basis and approach to developing the solutions of various types of computational problems is briefly described in this section. This discussion is not meant to be exhaustive or comprehensive, but rather introductory, in nature. Several alternative techniques are available for solving the various types of problems; the following discussion is in most cases confined to presenting an outline of one of the techniques.

4.2.1 Linear Algebraic Equations

It should be clear that systems of linear algebraic equations can range in size from very small (fewer than five equations) to very large (several hundreds), depending on the number of components and complexity of operations. For example, a system consisting of four components being separated in a distillation column containing five stages yields a system of 20 material balance equations. Typically, the system of equations is rearranged into the following matrix form:

$$[A]\cdot[X]=[B] \tag{4.15}$$

In this equation, $[X]$ is the column matrix of n variables; $[A]$, the $n \times n$ matrix of coefficients; and $[B]$, a column matrix of n function values.

The *Gauss elimination* technique for solving this system of equations involves progressive elimination of variables from the equations such that at the end only a single linear equation is obtained in one variable. The value of that variable is then obtained and back-substituted progressively into the equations in reverse order of elimination to obtain the values of the rest of the variables that satisfy equation 4.15. For example, if the system consists of n equations in variables $x_1, x_2, \ldots, x_n$, then the first step is elimination of variable x_1 from equations 2 to n using equation 1 to express x_1 in terms of the rest of variables. The result is a system of $n - 1$ equations in $n - 1$ variables $x_2, x_3, \ldots, x_n$. Repeating this procedure then allows us to eliminate variables x_2, x_3, and so on, until only an equation in x_n is left. The value of x_n is calculated, and reversing the calculations, values of x_{n-1}, $x_{n-2}, \ldots, x_1$ are obtained [4].

Iterative procedures offer an alternative to elimination techniques. The *Gauss-Seidel* method involves assuming an initial solution by guessing the values for the variables. It is often convenient to assume that all the variables are 0. Based on this initial guess, the values of the variables are recalculated using the system of equations: x_1 is calculated from the first equation, and its value is updated in the solution matrix; x_2 is calculated from the second equation; and so on. The steps are repeated until the values converge for each variable [8]. The Gauss-Seidel method is likely to be more efficient than the elimination method for systems containing a very large number of equations or systems of equations with a sparse coefficient matrix, that is, where the majority of coefficients are zero [9].

Many sophisticated variations of the elimination and iteration techniques are available for the solution. One other solution technique involves matrix inversion and multiplication. The effectiveness of these and solution techniques is dependent on the nature of the system of equations. Certain techniques may work better in some situations, whereas it might be appropriate to use alternative techniques in other instances.

4.2.2 Polynomial and Transcendental Equations

The complexity of solutions for polynomial and transcendental equations increases with increasing nonlinearity. Quadratic equations can be readily solved using the quadratic formula, provided such equations can be readily rearranged in the appropriate form. Formulas exist for obtaining roots of a cubic equation, but these are rarely used. No such easy formulas are available for solution of higher-order polynomials and transcendental equations.

These equations are typically solved by guessing a solution (root) and refining the value of the root on the basis of the behavior of the function. The principle of the *Newton-Raphson* technique, one of the most common techniques used for determining the roots of an equation, is represented by equation 4.16 [4]:

$$x_{n+1} = x_n - \frac{f(x_n)}{f'(x_n)} \tag{4.16}$$

Here, x_n and x_{n+1} are the old and new values of the root; $f(x_n)$ and $f'(x_n)$ are values of the function and its derivative, respectively, evaluated at the old root.

The calculations are repeated iteratively; that is, so long as the values of the roots do not converge, the new root is reset as the old root and a newer value of the root evaluated. It is obvious the new root will equal the old root when the function value is zero. In practice, the two values do not coincide exactly, but a tolerance value is defined for convergence. For example, the calculations may be stopped when the two values differ by less than 0.1% (or some other acceptable criteria).

The computations for this technique depend on not only the function value but also its behavior (derivative) at the root value. The initial guess is extremely important, as the search for the root proceeds on the basis of the function and derivative values at this point. Proper choice of the root will yield a quick solution, whereas an improper choice of the initial guess may lead to the failure of the technique.

The iterative successive substitution method can also be used to solve such equations [9]. The method involves rearranging the equation $f(x) = 0$ in the form $x = g(x)$. The iterative solution algorithm can then be represented by the following equation:

$$x_{i+1} = g(x_i) \tag{4.17}$$

Here, x_{i+1} is the new value of the root, which is calculated from the old value of the root x_i. Each successive value of x would be close to the actual solution of the equation. The key to the success of the method is in the proper rearrangement of the equations, as it is possible for the values to diverge away from the solution rather than toward a solution.

Finding the roots of polynomial equations presents a particular challenge. An nth-order polynomial will have n roots, which may or may not be distinct and may be real or complex. The solution technique described previously may be able to find only a single root, irrespective of the initial guess. The polynomial needs to be *deflated*—its order reduced by factoring out the root discovered—progressively to find all the n roots. It should be noted that in engineering applications, only one root may be of interest, the others needed only for mathematically complete solution. For example, the cubic equation of state may have only one real positive root for volume, and that is the only root of interest to the engineer. A complex or negative root, while mathematically correct as an answer, is not needed by the engineer.

4.2.3 Derivatives and Differential Equations

Some computational problems may involve calculating or obtaining derivatives of functions. Depending on the complexity of the function, it may not be possible to obtain an explicit analytical expression for the derivative. Similarly, some of the problems may involve obtaining the derivative from observed data. For example, an experiment conducted for determination of the kinetics of a reaction will yield concentration-time data. An alternative method of determining the rate constant for the reaction involves regressing the rate of the reaction as a function of concentration. The rate of the reaction is defined as $-\frac{dC_A}{dt}$; thus, the problem involves estimating the derivative from the concentration-time data. One of the numerical techniques for obtaining the derivative is represented by equation 4.17.

$$\frac{dC_A}{dt} = \frac{\Delta C_A}{\Delta t} = \frac{C_{A_{i+1}} - C_{A_i}}{t_{i+1} - t_i} \tag{4.18}$$

The subscripts refer to the time period. Thus, C_{Ai} is the concentration at time t_i, and so on. The derivative is approximated by the ratio of differences in the quantities. This formula is termed the *forward difference* formula, as the derivative at t_i is calculated using values at t_i and t_{i+1}. Similarly, there are *backward* and *central difference* formulas that are also applied for the calculation of the derivative [4, 10]. The comparative advantages and disadvantages of the different formulas are beyond the scope of this book and are not discussed further.

Similarly, the numerical techniques for integration of ordinary and partial differential equations are beyond the scope of this book. Interested readers may find a convenient starting point in reference [4] for further knowledge of such techniques.

4.2.4 Regression Analysis

The common basis for linear as well as multiple regression is the minimization of the sum of squared errors (SSE) between the experimentally observed values and the values predicted by the model, as shown in equation 4.19:

$$SSE = \sum_{i=1}^{n} \left(y_i - f(x_i)\right)^2 \tag{4.19}$$

In this equation, y_i is the observed value, and $f(x_i)$ is the predicted value based on the presumed function f. The function can be linear in a single variable (generally, what is implied by the term linear regression), linear in multiple variables (multiple regression), or polynomial (polynomial regression). Minimization of SSE yields values of model parameters (slope and intercept for a linear function, for example) in terms of the observed data points (x_i, y_i). The *least squares* regression formulas are built into many software programs.

4.2.5 Integration

As mentioned previously, numerical computation of an integral is needed when it is not possible to integrate the expression analytically. In other cases, discrete values of the function may be available at various points. Numerical integration of such functions involves summing up the weighted values of the function evaluated or observed at specified points. The fundamental approach is to construct a trapezoid between any two points, with the two parallel sides being the function values and the interval between the independent variable values constituting the height [4, 10]. If the function is evaluated at two points, a and b, then the following applies:

$$\int_a^b f(x)dx = \frac{b-a}{2} \cdot [f(b) - f(a)] \tag{4.20}$$

Decreasing the interval increases the accuracy of the estimate. Several other refinements are also possible but are not discussed here.

Section 4.3 describes various software programs that are available for the computations and solutions of the different types of problems just discussed. These software programs feature built-in tools developed on the basis of these algorithms, obviating any need for an engineer to write a detailed program customized for the problem at hand. The engineer has to know merely how to give the command in the language that is understood by the program. The previous discussion should, however, provide the theoretical basis for the solution as well as illustrate the limitations of the solution technique and possible causes of failure. A course in numerical techniques is often a required core course in graduate chemical engineering programs and sometimes an advanced undergraduate elective course.

4.3 Computational Tools—Machines and Software

The two components enabling the computations are the hardware—machines that perform the calculations—and the software—the instructions to run solution algorithms by the machines. Both these components are described in the sections that follow.

4.3.1 Computational Machines

A counting frame, or an *abacus*, is one of the earliest devices and has been used for more than three millennia for rapid calculations. This remarkable device continues to be used in various parts of the world to improve on manual hand calculations. Abaci are also popular as teaching tools in elementary schools. Calculations using an abacus involve physically moving beads on a wire, which clearly constrains the speed and type of calculations that can be performed.

A couple of tools developed in response to the ever-increasing demand for higher speeds worth mentioning are the *log table* and the *slide rule*. Figure 4.8 shows a page from the log table.

A log table enables all calculations, including highly complex ones, to be reduced to additions/subtractions, which can then be easily done by hand. The use of log tables is illustrated for the simple case of calculating the circumference C of a circle of diameter D. The concept and steps in calculations are as follows:

Equations 4.21, 4.22, and 4.23 demonstrate how a multiplication operation can be performed as an addition operation using the log tables. Logarithms of π and diameter are looked up in the log tables and added to obtain logarithm of the circumference. The answer is obtained by looking up the antilog of the sum in another set of tables.

$$C = \pi D \tag{4.21}$$

$$\log C = \log \pi + \log D \tag{4.22}$$

$$C = \text{Antilog}(\log C) \tag{4.23}$$

The calculation will work with logarithms to any base; however, generally the tables are those of common logarithms (to the base 10) rather than natural logarithms (to the base e). An impressive compilation of common logarithms up to 24 decimal places of numbers leading to 200,000 was available by the end of the 18th century through the efforts of a large number of individuals.

Such compilations and early calculating machines were hardly portable, a disadvantage that was overcome by the slide rule. This elegant device was the size of a ruler and small enough to fit in a shirt pocket; a couple of examples are shown in Figure 4.9. As seen from the figure, a slide rule has a central sliding part, and both the fixed and sliding parts are marked with several scales. Multiplication, division, and logarithmic as well as trigonometric calculations can be rapidly performed using the slide rule, which became an indispensable tool for engineers until it was supplanted by inexpensive scientific calculators in the mid-1970s.

LOGARITHMS

Natural Numbers	0	1	2	3	4	5	6	7	8	9	Proportional Parts								
											1	2	3	4	5	6	7	8	9
10	0000	0043	0086	0128	0170	0212	0253	0294	0334	0374	4	8	12	17	21	25	29	33	37
11	0414	0453	0492	0531	0569	0607	0645	0682	0719	0755	4	8	11	15	19	23	26	30	34
12	0792	0828	0864	0899	0934	0969	1004	1038	1072	1106	3	7	10	14	17	21	24	28	31
13	1139	1173	1206	1239	1271	1303	1335	1367	1399	1430	3	6	10	13	16	19	23	26	29
14	1461	1492	1523	1553	1584	1614	1644	1673	1703	1732	3	6	9	12	15	18	21	24	27
15	1761	1790	1818	1847	1875	1903	1931	1959	1987	2014	3	6	8	11	14	17	20	22	25
16	2041	2068	2095	2122	2148	2175	2201	2227	2253	2279	3	5	8	11	13	16	18	21	24
17	2304	2330	2355	2380	2405	2430	2455	2480	2504	2529	2	5	7	10	12	15	17	20	22
18	2553	2577	2601	2625	2648	2672	2695	2718	2742	2765	2	5	7	9	12	14	16	19	21
19	2788	2810	2833	2856	2878	2900	2923	2945	2967	2989	2	4	7	9	11	13	16	18	20
20	3010	3032	3054	3075	3096	3118	3139	3160	3181	3201	2	4	6	8	11	13	15	17	19
21	3222	3243	3263	3284	3304	3324	3345	3365	3385	3404	2	4	6	8	10	12	14	16	18
22	3424	3444	3464	3483	3502	3522	3541	3560	3579	3598	2	4	6	8	10	12	14	15	17
23	3617	3636	3655	3674	3692	3711	3729	3747	3766	3784	2	4	6	7	9	11	13	15	17
24	3802	3820	3838	3856	3874	3892	3909	3927	3945	3962	2	4	5	7	9	11	12	14	16
25	3979	3997	4014	4031	4048	4065	4082	4099	4116	4133	2	3	5	7	9	10	12	14	15
26	4150	4166	4183	4200	4216	4232	4249	4265	4281	4298	2	3	5	7	8	10	11	13	15
27	4314	4330	4346	4362	4378	4393	4409	4425	4440	4456	2	3	5	6	8	9	11	13	14
28	4472	4487	4502	4518	4533	4548	4564	4579	4594	4609	2	3	5	6	8	9	11	12	14
29	4624	4639	4654	4669	4683	4698	4713	4728	4742	4757	1	3	4	6	7	9	10	12	13
30	4771	4786	4800	4814	4829	4843	4857	4871	4886	4900	1	3	4	6	7	9	10	11	13
31	4914	4928	4942	4955	4969	4983	4997	5011	5024	5038	1	3	4	6	7	8	10	11	12
32	5051	5065	5079	5092	5105	5119	5132	5145	5159	5172	1	3	4	5	7	8	9	11	12
33	5185	5198	5211	5224	5237	5250	5263	5276	5289	5302	1	3	4	5	6	8	9	10	12
34	5315	5328	5340	5353	5366	5378	5391	5403	5416	5428	1	3	4	5	6	8	9	10	11
35	5441	5453	5465	5478	5490	5502	5514	5527	5539	5551	1	2	4	5	6	7	9	10	11
36	5563	5575	5587	5599	5611	5623	5635	5647	5658	5670	1	2	4	5	6	7	8	10	11
37	5682	5694	5705	5717	5729	5740	5752	5763	5775	5786	1	2	3	5	6	7	8	9	10
38	5798	5809	5821	5832	5843	5855	5866	5877	5888	5899	1	2	3	5	6	7	8	9	10
39	5911	5922	5933	5944	5955	5966	5977	5988	5999	6010	1	2	3	4	5	7	8	9	10
40	6021	6031	6042	6053	6064	6075	6085	6096	6107	6117	1	2	3	4	5	6	8	9	10
41	6128	6138	6149	6160	6170	6180	6191	6201	6212	6222	1	2	3	4	5	6	7	8	9
42	6232	6243	6253	6263	6274	6284	6294	6304	6314	6325	1	2	3	4	5	6	7	8	9
43	6335	6345	6355	6365	6375	6385	6395	6405	6415	6425	1	2	3	4	5	6	7	8	9
44	6435	6444	6454	6464	6474	6484	6493	6503	6513	6522	1	2	3	4	5	6	7	8	9
45	6532	6542	6551	6561	6571	6580	6590	6599	6609	6618	1	2	3	4	5	6	7	8	9
46	6628	6637	6646	6656	6665	6675	6684	6693	6702	6712	1	2	3	4	5	6	7	7	8
47	6721	6730	6739	6749	6758	6767	6776	6785	6794	6803	1	2	3	4	5	5	6	7	8
48	6812	6821	6830	6839	6848	6857	6866	6875	6884	6893	1	2	3	4	4	5	6	7	8
49	6902	6911	6920	6928	6937	6946	6955	6964	6972	6981	1	2	3	4	4	5	6	7	8
50	6990	6998	7007	7016	7024	7033	7042	7050	7059	7067	1	2	3	3	4	5	6	7	8
51	7076	7084	7093	7101	7110	7118	7126	7135	7143	7152	1	2	3	3	4	5	6	7	8
52	7160	7168	7177	7185	7193	7202	7210	7218	7226	7235	1	2	2	3	4	5	6	7	7
53	7243	7251	7259	7267	7275	7284	7292	7300	7308	7316	1	2	2	3	4	5	6	6	7
54	7324	7332	7340	7348	7356	7364	7372	7380	7388	7396	1	2	2	3	4	5	6	6	7

Figure 4.8 A log table page.

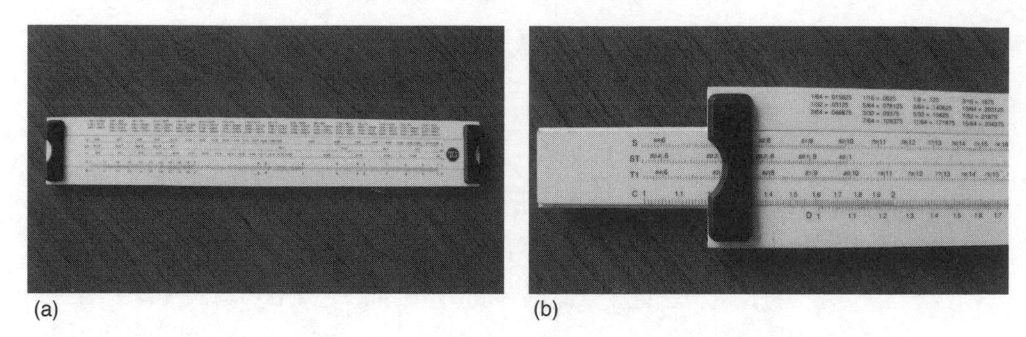

(a) (b)

Figure 4.9 Slide rule—(a) closed, (b) open (sliding scale pulled out).

The advent of computers enabled engineers to perform highly complex, involved, repetitive calculations and improve the accuracy of solutions. Mainframe computers were ubiquitous by the 1950s, with engineers writing programs and submitting them as jobs to be run on the mainframes.

Rapid technological advances in memory and data storage, materials, and processors resulted in lowering the cost of computers, making them affordable to most people. By the mid-1990s, personal computers (PCs) had become ubiquitous and an essential accessory for an engineering student. Continual reduction in the cost and size of devices has resulted in the availability of portable devices such as laptops and tablets that allow us to perform practically all types of computations except for the most complicated ones that require the use of supercomputers.

4.3.2 Software

As previously mentioned, computing machines need to be given instructions for performing calculations. These instructions are codified in a well-defined structure—the programming language. All the computational steps are written according to the rules of the programming language, resulting in a program, which can then be compiled and executed by the computer [4].

Developed in the 1950s, *Fortran* (from *for*mula *tran*slation) became the dominant programming language for scientific computing, and even today remains the preferred language for highly intensive computations of large systems. Each line of the Fortran program used to be transferred onto a card using a keypunch machine. (An IBM punched card with 80 columns is shown in Figure 4.10. Each number, letter, and symbol was represented by

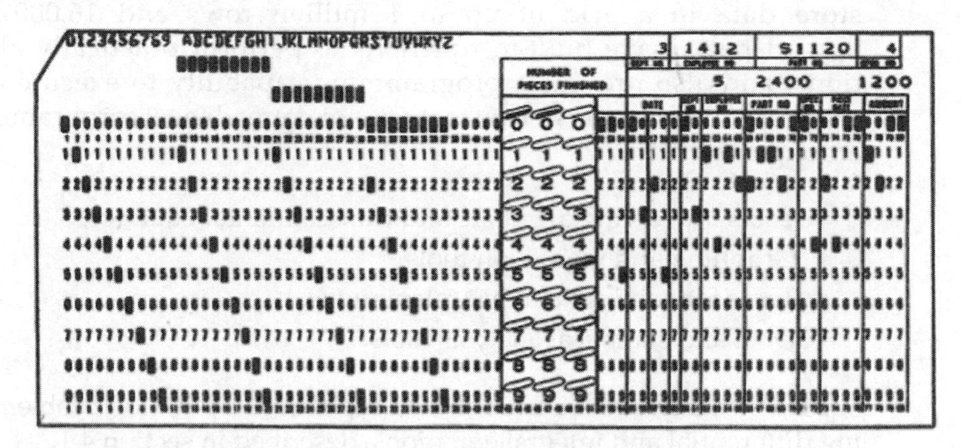

Figure 4.10 An IBM punched card illustrating representation of numbers and letters by punched holes.

Source: Ceruzzi, P. E., *A History of Modern Computing*, Second Edition, MIT Press, Cambridge, Massachusetts, 2003.

a hole or holes punched in the column [11]). A deck of such punched cards would constitute a program, which along with the necessary data would be submitted to be run as a job on the mainframe computer. Subsequent developments eliminated the need for punch cards, and a Fortran program can now be run on a personal computer, similar to programs in other languages.

A large number of program modules developed over the past 60 years have resulted in a valuable library of programs in Fortran, and these modules are used every day throughout scientific and engineering computations.

Although Fortran remains indispensable for advanced computing, a large number of advanced software programs are available on the PC to perform most of the chemical engineering computations. The ubiquitous availability of these software programs has essentially eliminated the need for individuals to write their own software codes for all but only very specific problems [12]. Following are descriptions of a few of these software packages.

4.3.2.1 Spreadsheets

A spreadsheet program is an integral component of a software package for PCs, with Microsoft Excel being the dominant one. Spreadsheets store data in a grid of up to 1 million rows and 16,000 columns. Spreadsheets have built-in functions to perform practically all calculations, and also provide a programming capability to execute any other type of computations. The features of spreadsheet programs include, among others,

- Tools for solving algebraic and transcendental equations
- Graphing and regression tools
- Tools for solving iterative calculations
- Data and statistical analysis tools

Spreadsheets can be used to solve all types of computation problems, including differential and integral equations, described in section 4.1.

4.3.2.2 Computing Packages

A number of software packages that provide a computing environment have become available over the past 30 years. These include MATLAB (Mathworks, Inc., Massachusetts, USA), Mathematica (Wolfram Research, Illinois, USA), Maple (Waterloo Maple, Ontario, Canada), Mathcad (MathSoft, Massachusetts, USA), and others. These packages typically offer, to varying degrees, an ability to perform numerical and symbolic computations, graphing, as well as programming. Each package has its characteristic *syntax*—rules for constructing instructions—and degree of user-friendliness. As with spreadsheets, practically all types of computational problems can be handled by these software packages that also provide an additional benefit of having built-in scientific/engineering constants and units. Appendix A presents a comparative discussion of some of these programs.

4.3.2.3 COMSOL

COMSOL (Comsol Group, Sweden and USA) is a powerful software package specifically geared to solve coupled physics and engineering problems. Marketed as a multiphysics software, COMSOL offers a modeling and simulation environment for all engineering disciplines. The various modules available in COMSOL allow users to solve problems related to fluid flow, heat transfer, reaction engineering, electrochemistry, and many others.

4.3.2.4 Process Simulation Software

Chemical process plants are complex structures consisting of a large number of units. The software systems previously described support computations related to a single step or a single unit. Several comprehensive process simulation software packages are used by the chemical industry to design, control, and optimize operations of chemical process plants. The computational power offered by these software systems enables a chemical engineer to obtain integrated plant design in a fraction of the time needed before their advent. Following are some prominent software examples:

- Aspen Plus by AspenTech, Massachusetts, USA
- PRO/II by Invensys, Texas, USA (a division of Schneider Electric)
- ProSimPlus by ProSim, S.A., France
- CHEMCAD by Chemstations, Inc., Texas, USA

All of these software systems offer similar capabilities and user interfaces. A design engineer will invariably be using one of these environments for designing an operation, a unit, or the plant. Appendix B illustrates the capabilities of one of these software programs—PRO/II—for solving a complex separation problem to provide an idea of the amazing computational power at our disposal.

In addition to these general-purpose software programs, some specialized software for specific applications is also available, such as FLOTRAN/ANSYS (ANSYS Inc., Pennsylvania, United States) for pipe flow/fluid dynamics computations. Open source software such as Modelica (JModelica.org) is also available for modeling and simulation of complex systems.

4.4 Summary

Chemical engineers perform a wide variety of computations ranging from simple arithmetic calculations to solving partial differential equations and simulating entire process plants. Technological advances have enabled engineers to access high-performance computing machines and use advanced software for obtaining solutions rapidly. As envisioned by Leibniz, this has allowed them the freedom from time-consuming, laborious, repetitive computational tasks, and they can focus on the more important mission of developing technologies from concept to practical implementation.

References

1. Varma, A., and M. Morbidelli, *Mathematical Methods in Chemical Engineering*, Oxford University Press, Oxford, England, 1997.
2. Wankat, P. C., *Separation Process Engineering*, Third Edition, Prentice Hall, Upper Saddle River, New Jersey, 2012.
3. Kyle, B. G., *Chemical and Process Thermodynamics*, Third Edition, Prentice Hall, Upper Saddle River, New Jersey, 1999.
4. Chapra, S. C., and R. P. Canale, *Numerical Methods for Engineers*, Seventh Edition, McGraw-Hill, New York, 2014.
5. Welty, J. R., C. E. Wicks, R. E. Wilson, and G. L. Rorrer, *Fundamentals of Momentum, Heat, and Mass Transfer*, Fifth Edition, John Wiley and Sons, New York, 2008.
6. Fogler, H. S., *Elements of Chemical Reaction Engineering*, Fourth Edition, Prentice Hall, Upper Saddle River, New Jersey, 2006.
7. Doraiswamy, L. K., and D. Üner, *Chemical Reaction Engineering: Beyond the Fundamentals*, CRC Press, Boca Raton, Florida, 2014.
8. Rice, R. G., and D. D. Do, *Applied Mathematics and Modeling for Chemical Engineers*, Second Edition, John Wiley and Sons, New York, 2012.
9. Riggs, J. B., *An Introduction to Numerical Methods for Chemical Engineers*, Second Edition, Texas Tech University Press, Lubbock, Texas, 1994.
10. Pozrikidis, C., *Numerical Computation in Science and Engineering*, Second Edition, Oxford University Press, New York, 2008.
11. Ceruzzi, P. E., *A History of Modern Computing*, Second Edition, MIT Press, Cambridge, Massachusetts, 2003.
12. Finlayson, B. A., *Introduction to Chemical Engineering Computing*, Second Edition, John Wiley and Sons, New York, 2014.

Problems

4.1 Cramer's rule and matrix inversion-multiplication offer alternative techniques to solve a system of linear algebraic equations. Conduct a literature search to collect information about these two techniques and the elimination and iteration techniques discussed in this chapter. Compare the various techniques regarding the complexity of algorithms, ease of implementation, and potential errors.

4.2 The Newton-Raphson technique may not converge to a solution. Inspecting equation 4.16, in what other possible way can the technique fail?

4.3 Roots of any equation can be found using what is known as the *bracketing technique*. Conduct a literature search and explain the principle behind such solution techniques.

4.4 The following data were obtained in an experiment where the concentration of a substance was monitored as a function of time. Calculate the first derivative of the concentration with respect to time for all possible times using the forward difference formula. Can the second derivative also be calculated numerically?

Time, s	**Concentration**
0	0
10	0.5
20	1.0
30	2.0
40	4.0
50	5.5
60	6.5
70	7.0
90	7.7

4.5 What is the area under the concentration-time curve obtained from the data shown for problem 4.4? Use the trapezoid method. An alternative technique is to use the rectangle method. What is the difference in the areas if the area is calculated using the rectangle method?

CHAPTER 5

Computations in Fluid Flow

Big whirls have little whirls
That feed on their velocity
And little whirls have lesser whirls
And so on to viscosity.

—Lewis Fry Richardson[1]

The landscape of a chemical process plant is dominated by a maze of pipes connecting various pieces of large equipment, transferring material from one point to another. Designing and optimizing this piping system requires a chemical engineer to have knowledge of the phenomena occurring in the flowing fluid. The fluid mechanics and transport phenomena courses explain these phenomena starting at the molecular level. This chapter presents general concepts regarding the nature of fluid flow and basic calculations associated with them.

5.1 Qualitative Description of Flow in Conduits

Consider a pipe closed at both ends and filled with a liquid. The liquid is composed of a very large number of molecules, each one of which occupies a certain position in the stagnant body of the liquid. The position of the molecule is not completely fixed, as it would be in the case of a solid, nor is it completely random, as in the case of a gas. A molecule may exhibit a slight random motion; however, it can be considered to be in a state of dynamic equilibrium—an overall fixed position with respect to time. Now, valves are opened at both ends of the pipe and the liquid is allowed to flow continuously from an upstream source.

At low flow rates, the molecules tend to follow a straight-line trajectory in the flow direction. The position of a molecule does not change with respect to that of the neighboring molecules in the direction of flow, and there is little or no tendency on the part of the molecule to move in a

1. An English mathematician known for his contributions in the field of weather forecasting and fractals. Quotation source: Richardson, L. F., *Weather Prediction by Numerical Process*, Second Edition, Cambridge University Press, Cambridge, England, 2007.

direction perpendicular to the direction of the flow. In other words, flow is strictly in the axial direction with no transverse movement of the molecules.

As the flow rate increases, this orderly arrangement tends to get broken. Molecules start deviating from their linear trajectories and begin to mix with other molecules in both the axial and transverse directions. As the flow rate increases still further, the movement of molecules becomes highly randomized, resulting in complete mixing in the transverse direction. The flow in the first case can be visualized as a flow of parallel layers with no intermixing of layers—a *laminar* flow—and the flow in the latter case can be visualized as completely mixed flow—a *turbulent* flow.

The two distinct flow patterns can be observed visually by injecting a colored dye in the flow stream. Figure 5.1 shows the evidence of laminar and turbulent flows obtained through such an experiment [1]. The left side of the figure shows the linear trajectory traced by the colored particles in the laminar flow, and the right side shows how the particles get dispersed in the body of the fluid in the turbulent flow. This experiment was originally performed in 1883 by Osborne Reynolds (whose influence on the field of fluid mechanics cannot be overstated) and can be easily set up in any laboratory.

5.1.1 Velocity Profiles in Laminar and Turbulent Flows

The distinction between laminar and turbulent flows becomes readily apparent from the velocity profiles—velocity of molecules or the particles as a function of position in the directions perpendicular to the flow direction, as shown in Figure 5.2 [2]. The velocity profile in laminar flow is parabolic, with zero velocity at the walls and maximum velocity at the center of the conduit. The velocity profile for turbulent flow, by contrast, is relatively flat, with most of the particles, except for those in a narrow region near the walls, having similar velocities.

The flow regime—laminar or turbulent—has a significant impact on the design and operation of the piping system. A chemical engineer

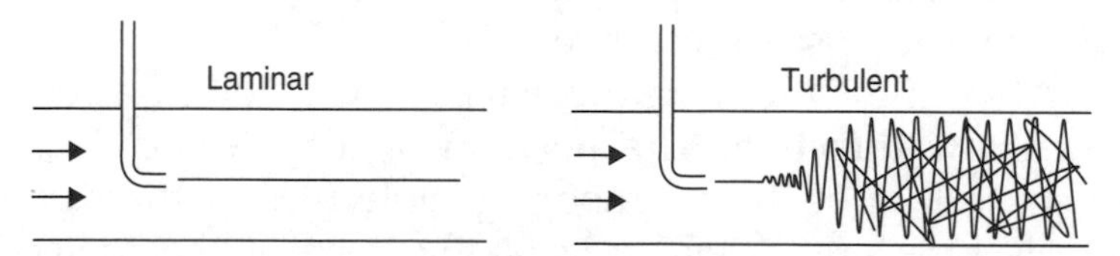

Figure 5.1 Laminar and turbulent flow.

Source: Olson, A. T., and K. A. Shelstad, *Introduction to Fluid Flow and the Transfer of Heat and Mass*, Prentice Hall, Englewood Cliffs, New Jersey, 1987.

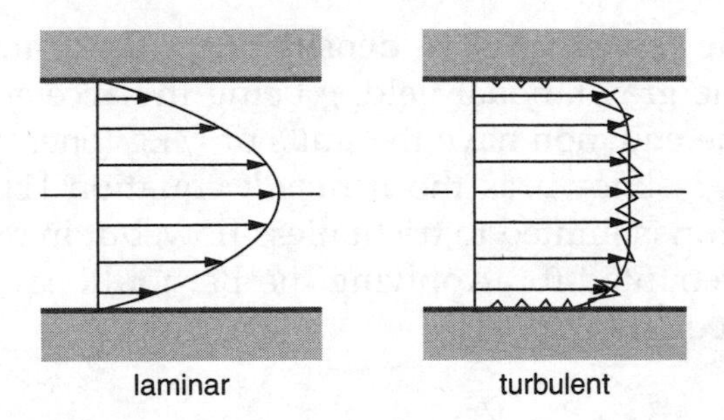

Figure 5.2 Velocity profiles in laminar and turbulent flows.
Source: Oertel, H., *Prandtl's Essentials of Fluid Mechanics*, Second Edition, Springer, New York, 2004.

designing such a system is required to specify the power requirements to move the material. It should be clear from this discussion that the phenomena occurring in the two regimes are fundamentally different. The power requirements have different dependence on system properties and operational conditions, and the engineer needs to have a quantitative criterion to determine if the flow is laminar or turbulent. The fundamental mathematical treatment of fluid flow phenomena is presented next.

5.2 Quantitative Analysis of Fluid Flow

As mentioned previously, one of the major responsibilities of a chemical engineer involves determination of power and energy requirements for the flow of fluids. This requires understanding the energy balance for the fluid flow, presented in section 5.2.1.

5.2.1 Energy Balance for Fluid Flow

The energy balance for systems involving a simple flow of a fluid is characterized by the lack of conversion of chemical, thermal, or any other kind of energy into mechanical energy. Essentially, the resulting mathematical formulation is simply a *mechanical energy balance* wherein the energy contributions arise from the potential and kinetic energy terms and flow work [3]. For an incompressible (constant density) fluid that does not experience any friction, the mechanical energy balance is given by equation 5.1 [1]:

$$\frac{P}{\rho}+\frac{V^2}{2}+gz=\text{constant} \tag{5.1}$$

where P is the pressure, V is the velocity, ρ is the density of the fluid, and z is the elevation of the fluid in the gravitational field, g being the acceleration due to gravity. The terms in the equation have the units of J/kg (energy per unit mass), and this equation is known as the *Bernoulli* equation [1]. The validity of the Bernoulli equation is limited to frictionless flow, but in reality, frictional effects need to be accounted for. Applying the Bernoulli equation between points 1 and 2 yields equation 5.2 [1]:

$$\frac{P_1}{\rho}+\frac{V_1^2}{2}+gz_1+w_{in}=\frac{P_2}{\rho}+\frac{V_2^2}{2}+gz_2+h_f \tag{5.2}$$

where h_f represents the frictional losses. It is assumed that the fluid flow is accomplished by means of a pump between the two points, and w_{in} is simply the work input from this pump. It should be clear from equation 5.2 that the power requirement for transferring a fluid from one point to another depends on the following four factors:[2]

- The difference in the hydrostatic pressure between the two points
- The difference in the elevation of the two points
- The difference between the fluid velocities at the two points
- The frictional losses in the piping system

The frictional losses depend on the flow regime, and the accounting of the frictional losses requires an understanding of the concept of viscosity.

5.2.2 Viscosity

The concept of friction is easily understood for a rigid, solid object. For a solid object to move, it must overcome the resistance from another object in contact with it. A similar situation can be envisioned in the case of a flowing fluid. Consider the velocity profile for laminar flow shown in Figure 5.2. The molecules are flowing in layers stacked upon each other. The molecules in contact with the wall can be considered to be stationary because of friction with the wall—a state called the *no-slip* condition. The layer of molecules adjoining this layer has to slide against this stationary layer. Similarly, there is relative motion between all adjacent layers, causing frictional losses. The resistance to motion between the layers is highest at the wall, decreasing toward the center of the conduit.

2. A student will encounter the detailed analysis while conducting the mechanical energy balance in later curriculum.

This variation in axial velocity with the position in the direction normal to the flow direction causes a shear stress in the fluid. This shear stress multiplied by the area over which it enacts yields the shear force—the frictional resistance to the flow. The shear stress for a fluid depends on the velocity gradient—how rapidly the velocity changes with distance. The following shows this mathematically [4]:

$$\tau = -\mu\left(\frac{dv_x}{dy}\right) \tag{5.3}$$

where τ is the shear stress, and v_x the velocity in the x-direction, which depends on the position in the y-direction. Equation 5.3, known as *Newton's law of viscosity*, indicates that the shear stress is directly proportional to the velocity gradient, with the constant of proportionality being μ, which is the *dynamic viscosity* of the fluid. Fluids that follow this simple relationship are termed *Newtonian fluids*. Many other fluids not conforming to this relationship are termed *non-Newtonian*. The negative sign in the equation arises from the directional considerations, with velocity and position being vector quantities, viscosity being a scalar, and the shear stress a tensor. The directional considerations are disregarded in the calculations in this book, and only absolute values are considered, reducing the equation to the following form:

$$|\tau| = \mu\left|\left(\frac{dv_x}{dy}\right)\right| \tag{5.4}$$

The SI unit of viscosity is $N{\cdot}s/m^2$ (equivalently, Pa·s or kg/m·s), the centimeter-gram-second (CGS) system unit being *poise* (P or g/cm·s). Viscosity is a function of temperature, the viscosity of water being 1 cP or 1 mPa·s at ~21°C. In contrast, viscosities of honey and glycerin are between 1500 and 2000 cP at ambient temperatures. The difficulty with which these fluids flow can be explained on the basis of their viscosities. Viscosities of substances can be predicted from theory or experimentally determined.

The force needed for getting the liquid to flow can be obtained by multiplying the shear stress with the area over which it acts, which is the contact surface area between the layers. The pressure differential needed to make the fluid flow can then be obtained by dividing the force by the cross-sectional flow area, which is the area in the direction perpendicular to the flow direction.

5.2.3 Reynolds Number

As previously mentioned, the flow regime is laminar at low flow rates. Frictional viscous forces predominate at these flow rates. As the flow rate increases, the orderly laminar arrangement is disrupted and the inertial forces associated with the movement of material begin to predominate. The ratio of these two forces can be related to the intrinsic fluid properties (viscosity μ, density ρ) and flow parameters (characteristics length dimension l, average velocity v) through a dimensionless quantity termed *Reynolds number* in honor of Osborne Reynolds:

$$Re = \frac{lv\rho}{\mu} \tag{5.5}$$

Reynolds, in his experiments, observed that the transition from laminar to turbulent flow occurred at the Reynolds number of 2300; that is, the flow was laminar below this value and began transitioning into turbulent flow above it. This transition may take place over a range of Reynolds numbers, the flow often considered to be fully turbulent when Re exceeds 4000. The characteristic length dimension depends on the system geometry. For cylindrical conduits, diameter d is used as the length dimension for calculating Re. It should be noted that these Re ranges are applicable to internal flows, that is, a flow through conduits. For external flows, that is, flows over surfaces and objects, such as flow around a car or an airplane, other quantitative criteria apply for the laminar to turbulent transition [5].

5.2.4 Pressure Drop Across a Flow Conduit

The frictional losses due to the flow of fluids result in a decrease of pressure from the point upstream to the point downstream. In other words, a higher upstream pressure is needed to overcome frictional losses in order to transfer fluid from the point upstream to the point downstream. The pressure drop can be viewed as the potential that induces fluid flow, analogous to voltage in an electrical circuit. Different mathematical expressions are used to obtain this pressure drop for the two different flow regimes.

The following equation shows the pressure drop for laminar flow through a pipe with constant circular cross section of diameter d:

$$\Delta P = \frac{128\mu LQ}{\pi d^4} \tag{5.6}$$

In this equation, ΔP is the pressure drop over pipe length L when the volumetric flow rate is Q. Equation 5.6 is called the *Hagen-Poiseuille equation* [1] in honor of G. Hagen and J. L. Poiseuille, who developed this formulation.

For turbulent flow, the pressure drop is given by the following:

$$\Delta P = \frac{2 f v^2 L \rho}{d} \tag{5.7}$$

where f is the friction factor[3], which is a function of the Reynolds number. Figure 5.3 shows the generalized trend exhibited by the friction factor as a function of the Reynolds number [3]. The friction factor, in the turbulent regimes, also shows a dependence on the roughness of the pipe, as indicated by the value of the parameter k/D in the figure.

Chemical engineers often use the Nikuradse equation (covered in Chapter 4, "Introduction to Computations in Chemical Engineering") for calculating the friction factor for turbulent flow through smooth pipes:

$$\frac{1}{\sqrt{f}} = 4.0 \log_{10}\left\{Re\sqrt{f}\right\} - 0.40 \tag{5.8}$$

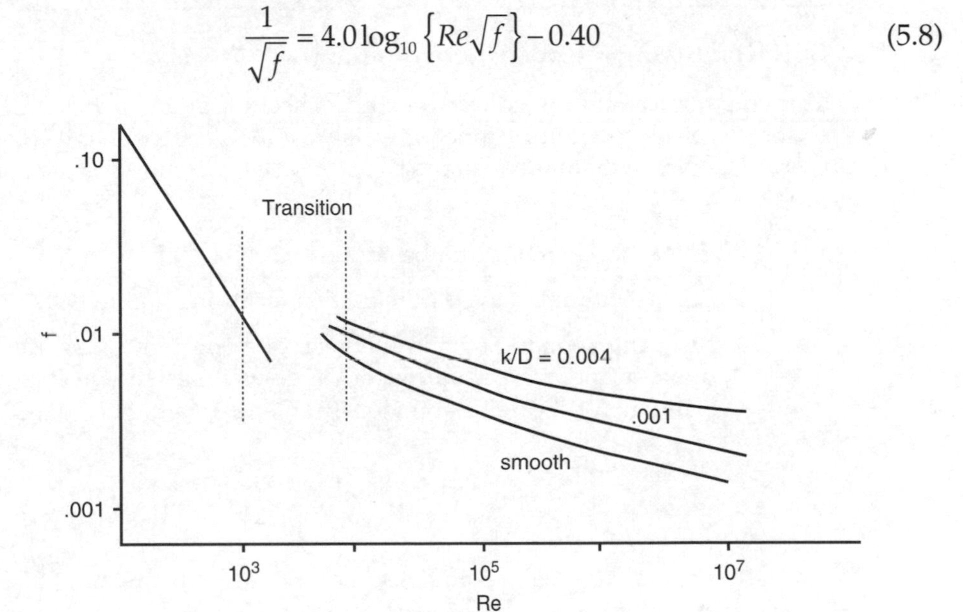

Figure 5.3 A generalized friction factor plot.
Source: Thomson, W. J., *Introduction to Transport Phenomena*, Prentice Hall, Upper Saddle River, New Jersey, 2000.

3. The friction factor, as used here and by chemical engineers generally, is the *Fanning friction factor*. Mechanical and civil engineers often use the *Darcy friction factor*, which, while calculated differently, has the same physical significance.

This equation is valid for flows having *Re* between 4000 and 3.2×10^6. The pressure drop so obtained is used further in the mechanical energy balance equation to calculate the power requirements for pumping the fluid.

5.3 Basic Computational Problems

Chemical engineers encounter problems ranging in complexity from simple arithmetic calculations to highly involved ones requiring programming. The following examples present of a few of these problems along with solution techniques using Excel and Mathcad.

Example 5.1 Friction Factor for Pipe Flow

Calculate the Fanning friction factor using the Nikuradse equation for the flow of water through a 1 in. diameter pipe when the Reynolds number is 10,000.

Solution (using a spreadsheet program)

This equation can be solved using a spreadsheet program such as Microsoft Excel. Excel has a built-in function called *Goal Seek* that enables users to solve the transcendental equation. The stepwise procedure for the solution follows:

1. Enter the Reynolds number value in a cell (B2).
2. Enter an initial guess for the friction factor in another cell (B3).
3. Rearrange equation 5.6 in the form of a function of the Reynolds number and the friction factor ($f(Re,f) = 0$), as shown below, and evaluate for the values previously entered in another cell (C3).

$$\frac{1}{\sqrt{f}} - 4.0\log\left\{Re\sqrt{f}\right\} + 0.40 = 0 \tag{E5.1}$$

4. The function value will most likely not equal 0, meaning the initial guess of friction factor was incorrect. Here, an initial guess of 0.1 for the friction factor leads to a function value of –10.44, indicating that the guess was incorrect. Click the DATA tab from the command menu, then the *What-if Analysis*, and select *Goal Seek*, as shown in Figure 5.4.[4]

4. Excel also has a much more powerful *Solver* utility, which allows a user to manipulate multiple cells, as well as specify constraints on the solution (for example, no negative roots or maximum/minimum solution permitted, etc.) for solving such problems.

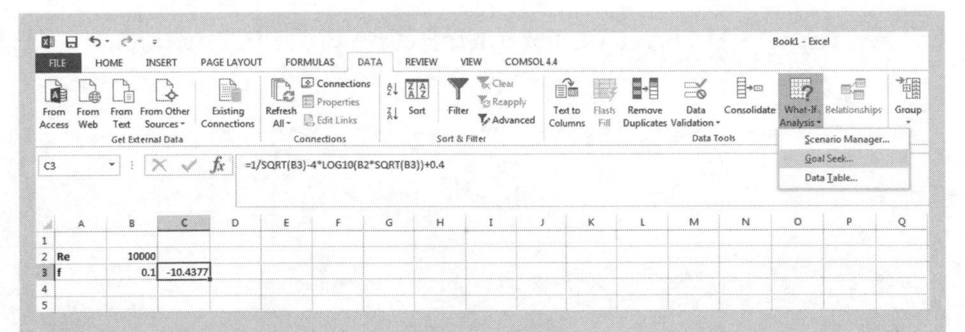

Figure 5.4 Goal Seek tool in Excel.

5. This brings up a dialog box, where cell C3 (function) is specified to be set to a value of 0 by manipulating cell B3 (friction factor f), as shown in Figure 5.5. Click OK.

Excel reaches a solution by changing the value in the cell B3, ultimately arriving at the f value of 0.0007727, with the function value in cell C3 reaching a value of –5.1E–6 (~0), as shown in Figure 5.6.

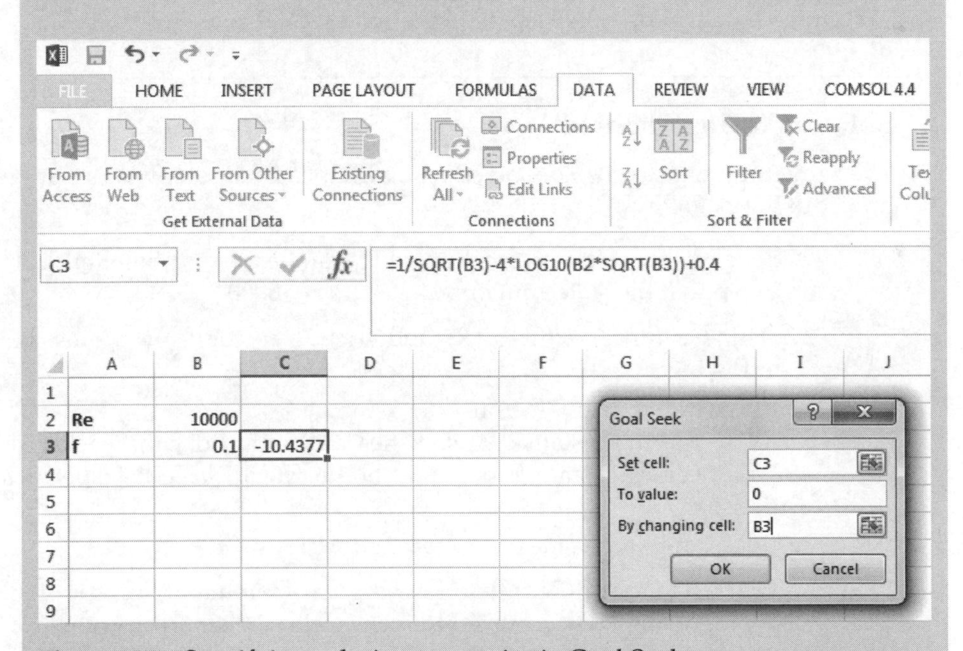

Figure 5.5 Specifying solution constraint in Goal Seek.

(Continues)

Example 5.1 Friction Factor for Pipe Flow (*Continued*)

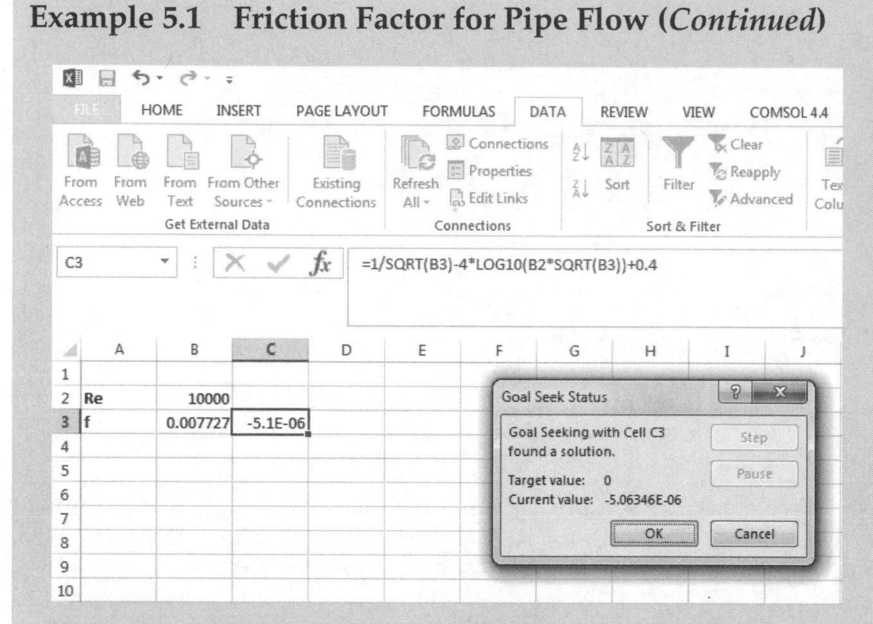

Figure 5.6 Solution of Nikuradse equation by Goal Seek in Excel.

Solution (using Mathcad)

The same problem can also be solved using Mathcad. The stepwise solution procedure is as follows:

1. Specify the value of the Reynolds number by typing Re:10000. (Note that the statement appears as Re := 10000.)
2. Define a function (ff) of *Re* and friction factor *f*, by typing the following:

   ff(Re,f):4.0*log(Re*\f)-0.40-1/\f

   (Note that the asterisk [*] specifies multiplication, the backslash [\] specifies square root, and the forward slash [/] specifies division.)
3. Guess an initial value of *f* by typing f:0.01.
4. Type f.ans:root(ff(Re,f),f) to use the *root* command for solution. Typing = at the end of this statement yields 7.727×10^{-3} as the friction factor, as shown in Figure 5.7.

The appearance of the statements is different from the characters typed. It should also be noted that there are alternative ways to specify the square root in Mathcad. The reader is left to explore the alternatives.

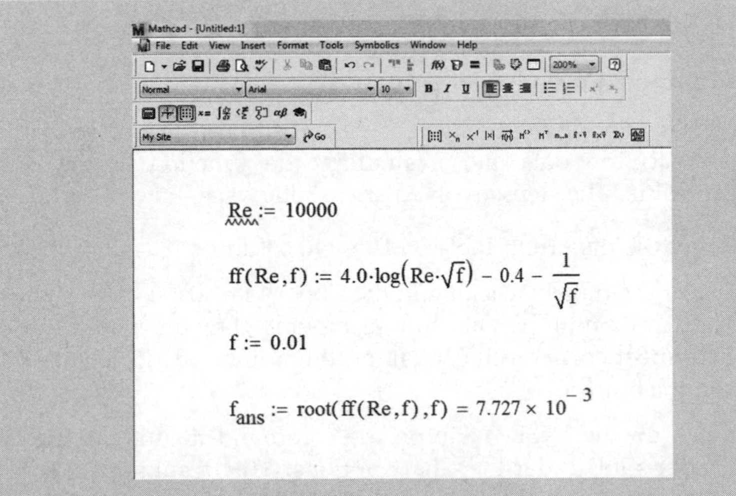

Figure 5.7 Solution of Nikuradse equation using Mathcad.

Similar procedures can be used to obtain the solution using other software—MATLAB, Mathematica, Maple, and so on—mentioned previously.

Example 5.2 Viscosity of a Fluid

The viscosity of a fluid is a function of temperature and fluid density. The experimentally measured viscosity of water at various temperatures is as shown in Table E5.1. What is the viscosity at 25°C?

Table E5.1 Water Viscosity as a Function of Temperature

T, °C	μ, mPa s
5	1.519
10	1.307
20	1.002
30	0.798
40	0.653
50	0.547
60	0.467
70	0.404
80	0.355
90	0.315

(Continues)

Example 5.2 Viscosity of a Fluid (*Continued*)

Solution (using a spreadsheet program)

The approach to solving this problem involves fitting a function (polynomial in this case) to the data and evaluating the function at the desired temperature value. The steps involved are as follows:

1. Enter the data from Table E5.1 in two columns.
2. Select the data by clicking in the upper leftmost corner of the data (cell containing number 5) and moving the cursor to the bottom rightmost corner (cell containing the number 0.315) before releasing the button.
3. Click on the INSERT tab on the command menu, and select the scatter plot under the Chart options. This results in an x-y plot being created from the data, as shown in Figure 5.8.

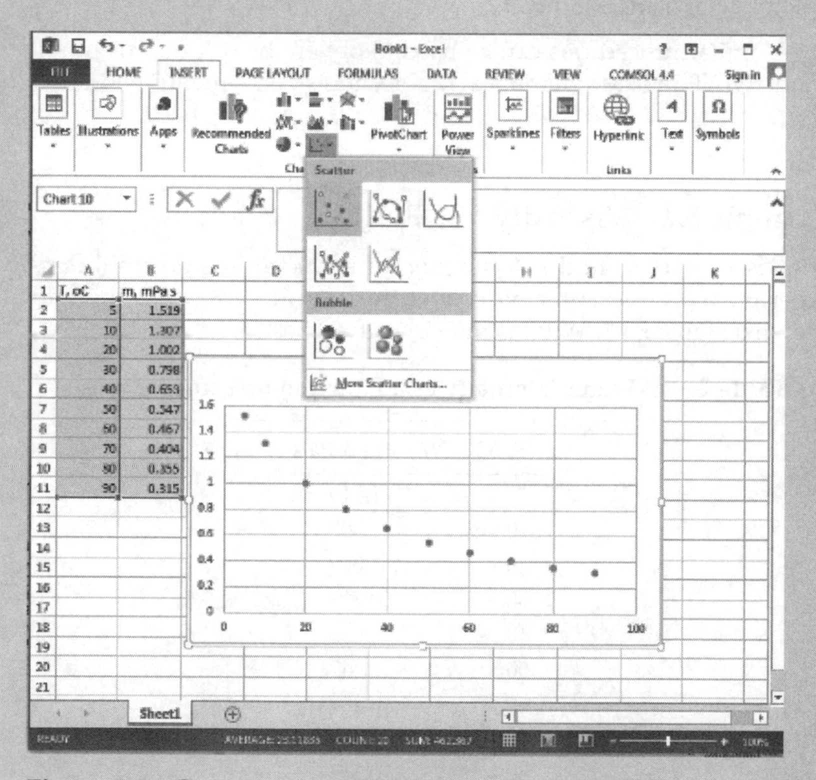

Figure 5.8 Creating a viscosity-temperature plot in Excel.

4. To fit a curve to the data, right-click on a data point to bring up an options box, and select Add Trendline from the menu, as shown in Figure 5.9.

5. A Format Trendline options box opens up to the right of the cells, allowing a number of options for the type of function to be fitted. Select Polynomial of third order (a cubic equation) and check the boxes to obtain the equation and correlation coefficient R^2 on the graph. The resulting equation is $y = -3e - 6x^3 + 0.0006x^2 - 0.045x + 1.7154$, y being the viscosity and x the temperature. The correlation

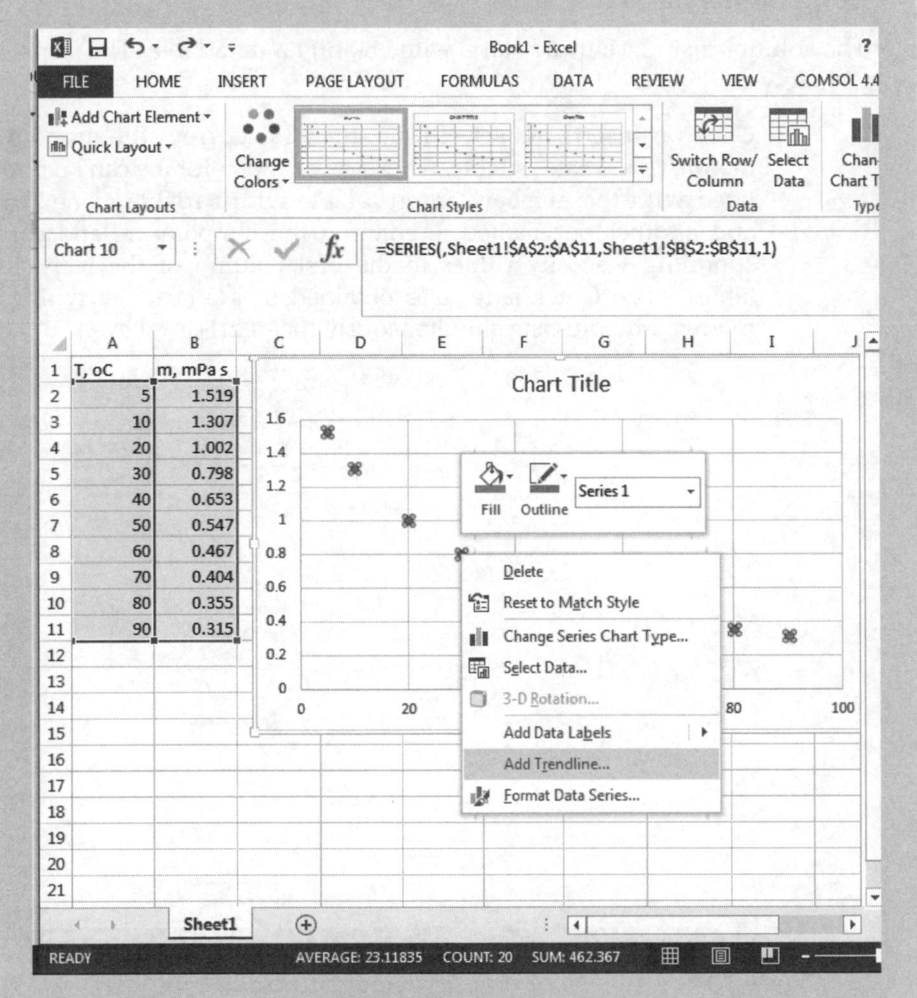

Figure 5.9 Fitting a trendline to the data.

(Continues)

Example 5.2 Viscosity of a Fluid (*Continued*)

coefficient for the fit is 0.9993, indicating that a third-order polynomial fits the data accurately within the given temperature range. The resulting graph, the equation of the trendline, and the correlation coefficient are shown in Figure 5.10.

6. Evaluate the function at $x = 25$ to calculate the viscosity at 25°C. This yields a viscosity of 0.9185 mPa s.

Solution (using Mathcad)

The solution using Mathcad starts with creating a data table. The steps are as follows:

1. Click on Insert, and select Data, then Table, from the dropdown menu. This creates a table, whose rows and columns can be populated with the numbers. Insert a table with variable name Temp and another one named μ, entering the temperature and corresponding viscosity values in the first column of the respective tables. (The Greek letter μ is obtained in Mathcad by typing *m* followed by pressing simultaneously the <Ctrl+g>[5] keys.)

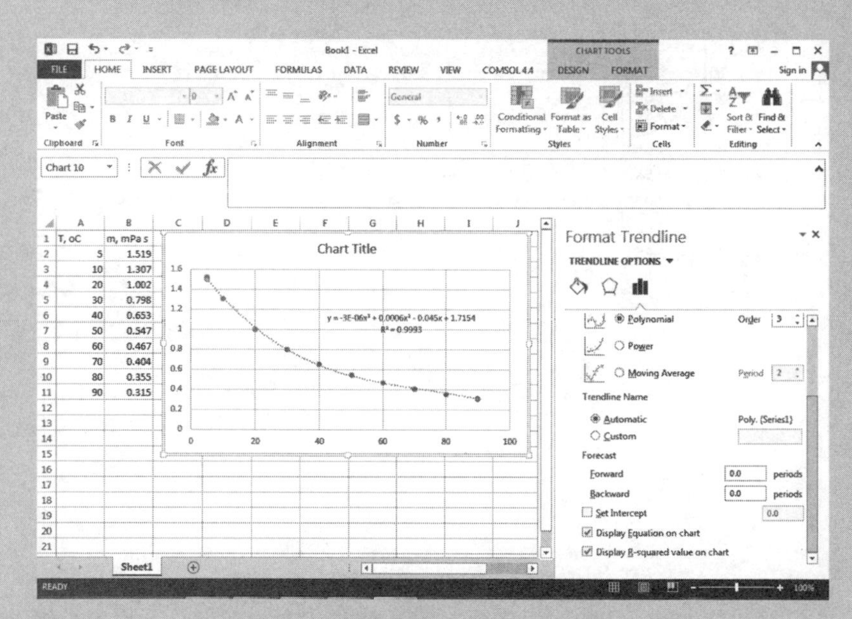

Figure 5.10 Cubic fit of viscosity-temperature data.

---

5. In this book, Ctrl key combinations are enclosed in angle brackets to indicate that the keys must be pressed simultaneously. For example, <Ctrl+g> means to hold down the Ctrl key and press g.

2. A regress function is used for polynomial regression. The arguments of the function are Temp, μ, and 3, representing the independent variable, dependent variable, and the order of the polynomial, respectively. Type vs:regress(Temp,μ,3)= to obtain a column vector, which consists of the following values: 3, 3, 3, 1.715, −0.045, 5.558×10^{-4}, and -2.556×10^{-6}, as shown in Figure 5.11. The first three numbers represent regress function, location of the first coefficient, and the order of the polynomial, and the last four numbers represent the values of the coefficients in the polynomial starting with the constant. The resulting equation is $\mu = -2.556e - 6\text{Temp}^3 + 0.000558\text{Temp}^2 - 0.045\text{Temp} + 1.715$, which is close to that obtained using Excel.

3. Use the interp function to obtain viscosity at 25°C. This function has four arguments: vs, Temp, μ, and 25; that is, the regression

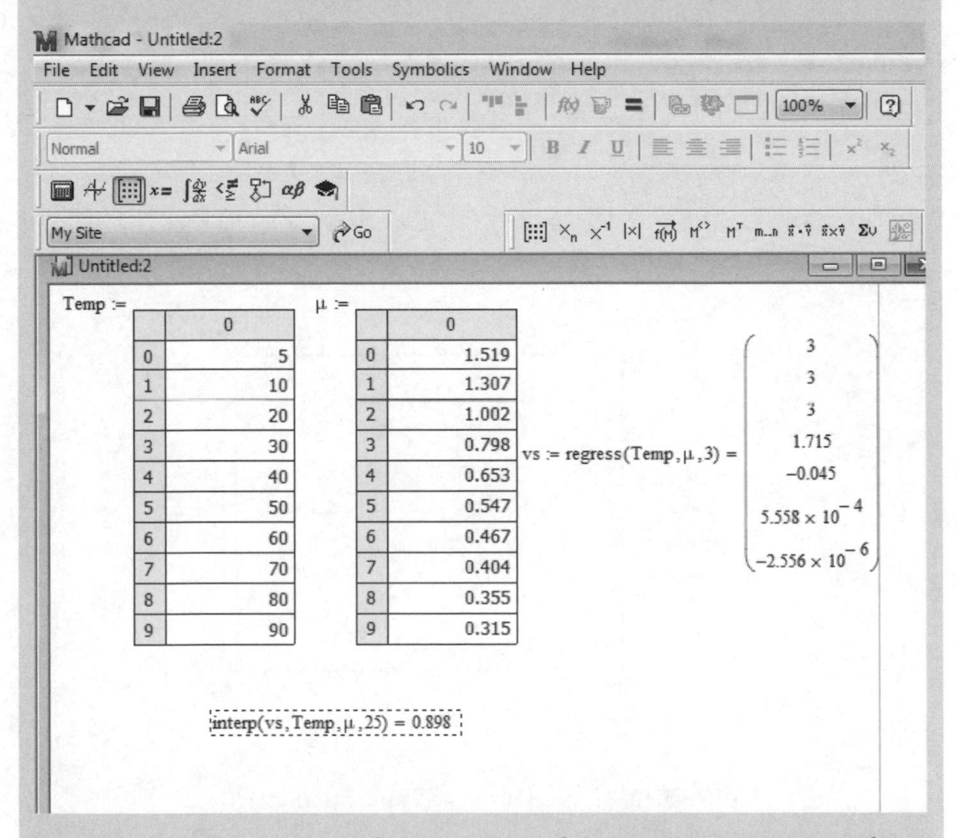

Figure 5.11 Regressing a cubic equation to the viscosity-temperature data in Mathcad.

(Continues)

Example 5.2 Viscosity of a Fluid (*Continued*)

vector, independent variable, dependent variable, and the value of the independent variable at which the function value is desired. Typing interp(vs,Temp,μ,25)= yields a value of 0.898 mPa s for the viscosity.

4. Define a viscosity function using the coefficients previously obtained by typing vis(T):-2.556*10^-6*T^3+5.558*10^-4*T^2-0.045*T+1.715.
5. To create a plot of observed and calculated viscosities, first define a range of temperatures: T:5,5.1;90. Click on the Graph Toolbar button, or choose Insert from the command menu followed by Graph in the dropdown menu, to insert a chart in the program. The resultant graph showing observed data as symbols and calculated values as a line is presented in Figure 5.12. The details of the manipulations are left to the reader.

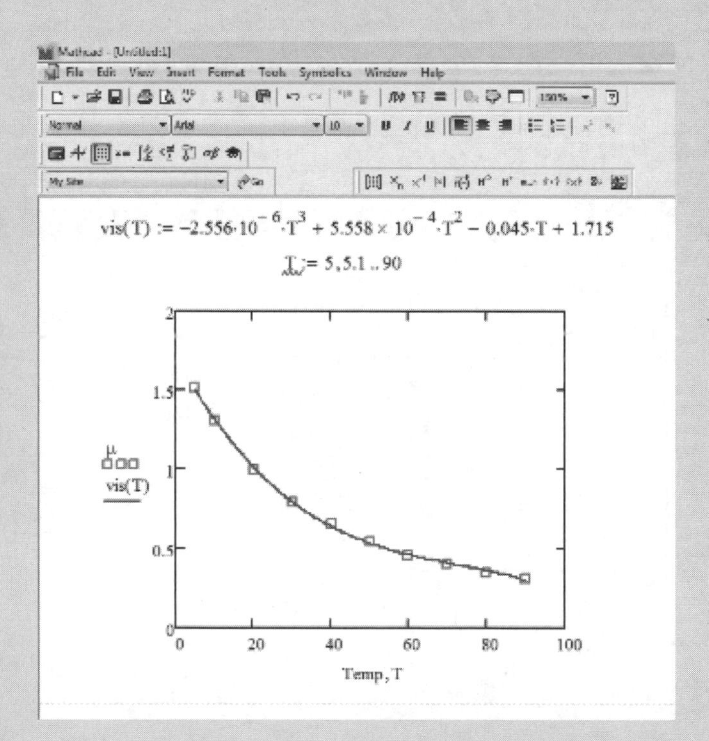

Figure 5.12 Plot of observed and calculated viscosities.

Example 5.3 Flow Rate, Average Velocity, and Reynolds Number

The velocity of a fluid flowing in a circular pipe was measured[6] as a function of the radial position and the resultant data are shown in Table E5.2. Calculate the flow rate, average velocity, and Reynolds number if the fluid is water at 25°C. Is the flow laminar?

Solution algorithm

The velocity across a cross section of the pipe is assumed to be a function of the radial position only; that is, if one draws a circle having that radius, the velocity is the same everywhere on that circle. In other words, the velocity does not depend on the *angular position* on that circle. Let $v(r)$ represent the fluid velocity through a thin ring of thickness Δr at radial location r. Then the following gives the flow rate ΔQ through this ring:

$$\Delta Q(r) = 2\pi r \Delta r v(r) \tag{E5.2}$$

Here, $2\pi r \Delta r$ is the area of the circular ring. The schematic representation of the situation is shown in Figure 5.13.

The total flow rate is simply the summation of all such flow rates calculated at the locations where the velocity is measured:

$$Q = \sum \Delta Q(r) = \sum 2\pi r \Delta r v(r) \tag{E5.3}$$

When the thickness of the ring becomes vanishingly small, equation E5.3 turns into an integral:

$$Q = \int_0^R 2\pi r v(r)\, dr \tag{E5.4}$$

Table E5.2 Velocity-Radial Position Data

Position, cm	Velocity, cm/s	Position, cm	Velocity cm/s	Position, cm	Velocity cm/s
0 (center)	10	0.4	8.4	0.8	3.6
0.1	9.9	0.5	7.5	0.9	1.9
0.2	9.6	0.6	6.4	1.0 (wall)	0.0
0.3	9.1	0.7	5.1		

(*Continues*)

6. There are several measurement techniques available for measuring the flow rate of a fluid in conduits. Measuring local velocities at various locations in the cross-sectional area for the flow using instruments such as a Pitot tube is one of these techniques.

Example 5.3 Flow Rate, Average Velocity, and Reynolds Number (*Continued*)

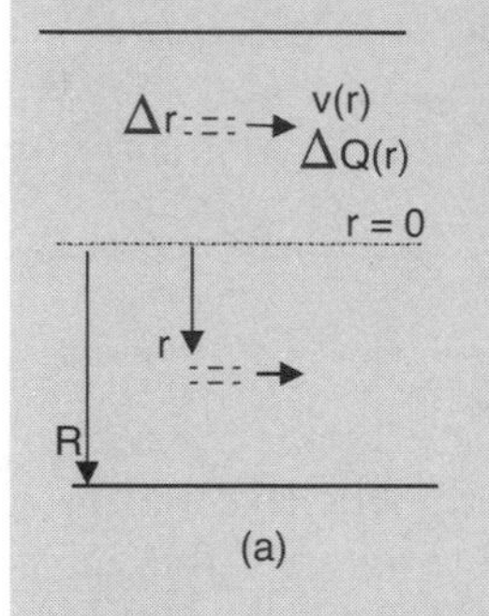

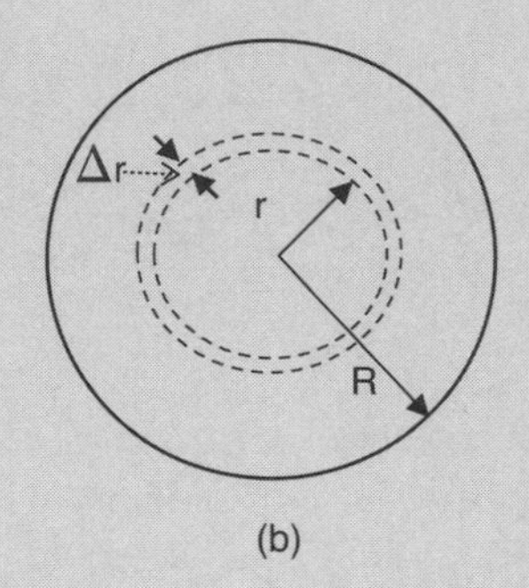

Figure 5.13 Flow through pipe—(a) side view, (b) cross-sectional view.

The average velocity and Reynolds number are calculated from equations E5.5 and E5.6, respectively.

$$v_{avg} = \frac{Q}{\frac{\pi}{4}D^4} \tag{E5.5}$$

$$Re = \frac{D v_{avg} \rho}{\mu} \tag{E5.6}$$

Here, D is the pipe diameter, ρ the density, and μ the viscosity of the fluid.

Solution (using a spreadsheet program)

The calculations using Excel are shown in Figure 5.14.

As seen from the figure, data from Table E5.2 are entered in columns A and B, and the flow rate at various locations is calculated in column C. The calculation for cell C4 can be seen in the figure, the formula entered by typing =2*pi()*A4*(A5-A4)*B4, corresponding to the terms in equation E5.2. The total flow rate is obtained by summing the flow rates in cells C3 through C12 by typing the formula =Sum(C3..C12) in cell C15. As seen from the figure, the total flow rate is calculated to be 15.55 cm^3/s. The average velocity is calculated by entering =C15/F7 in cell C16, and the Reynolds number by typing = C16*F8*F4/F5 in cell C17. The parameter values of the radius of the pipe, density, and viscosity are entered in cells F3, F4, and F5. The cross-sectional area is calculated in cell F7 (=pi()*F3^2), and cell F8 contains the diameter of the pipe, which is twice the radius (=2*F3). The average velocity is 4.95 cm/s, and the Reynolds number is 1100, indicating that the flow is laminar.

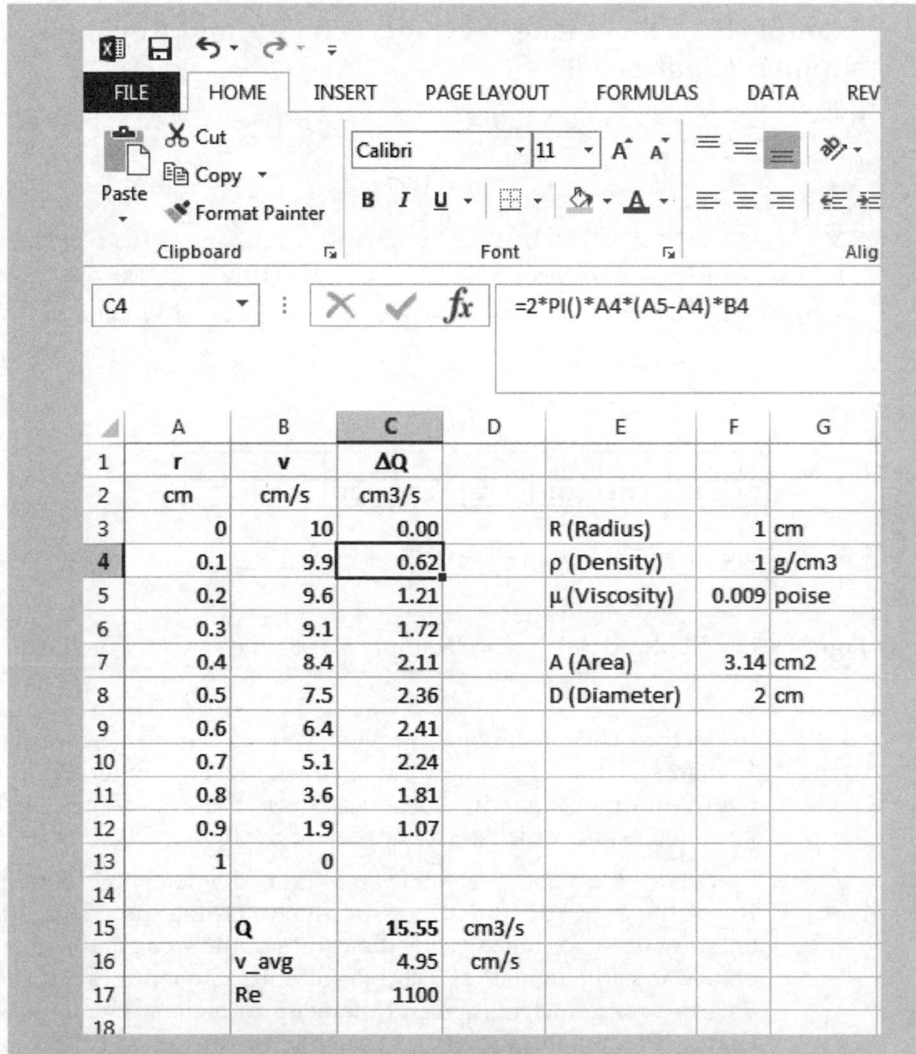

	A	B	C	D	E	F	G
1	r	v	ΔQ				
2	cm	cm/s	cm3/s				
3	0	10	0.00		R (Radius)	1	cm
4	0.1	9.9	0.62		ρ (Density)	1	g/cm3
5	0.2	9.6	1.21		μ (Viscosity)	0.009	poise
6	0.3	9.1	1.72				
7	0.4	8.4	2.11		A (Area)	3.14	cm2
8	0.5	7.5	2.36		D (Diameter)	2	cm
9	0.6	6.4	2.41				
10	0.7	5.1	2.24				
11	0.8	3.6	1.81				
12	0.9	1.9	1.07				
13	1	0					
14							
15		Q	15.55	cm3/s			
16		v_avg	4.95	cm/s			
17		Re	1100				
18							

Figure 5.14 Excel solution to Example 5.3.

Solution (using Mathcad)

The solution using Mathcad is shown in Figure 5.15, and the calculation steps follow.

1. Enter the parameters radius, density, and viscosity as follows: R:1*cm, r<Ctrl+g>:1*gm/cm^3, and m<Ctrl+g>:0.009*poise,

(Continues)

Example 5.3 Flow Rate, Average Velocity, and Reynolds Number (*Continued*)

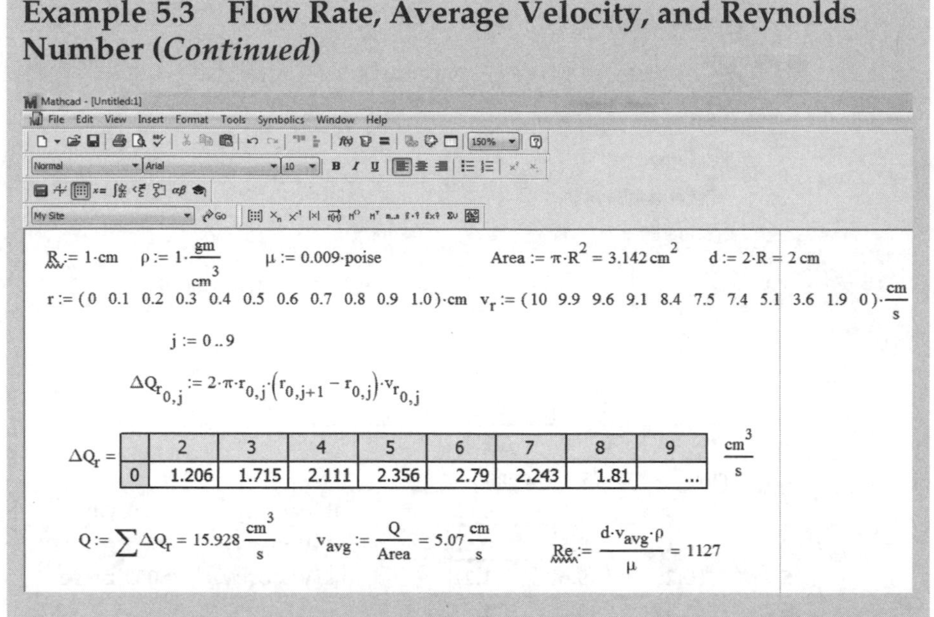

Figure 5.15 Mathcad Solution to Example 5.3.

respectively. (As mentioned previously, pressing <Ctrl+g> after a Roman script letter changes it to a Greek symbol.) Enter the units using multiplication/division operations. The appearance of the parameters and variables can be seen in the figure.

2. Calculate the cross-sectional area and diameter by entering A:p<Ctrl+g>*R^2 and d:2*R, respectively. Typing the = sign after entering these formulae yields the values of the area and the diameter. Mathcad automatically assigns SI units to quantities, yielding area in square meters (m^2) and diameter in meters. The units are changed to square centimeters (cm^2) and centimeters by clicking on the Mathcad units and entering the desired units.
3. Enter the radial position and velocity data by defining row matrices for each: r:<Ctrl+m> and v.r:<Ctrl+m>. The <Ctrl+m> key sequence brings up an Insert Matrix dialog box; enter the number of rows (1) and the number of columns (11). Click OK in the dialog box to create the matrix, and enter the respective data values. Assign units to the variables r and v_r in a similar manner as earlier; that is, using multiplication/division operations. (It should be noted that typing v.r creates a variable named v_r. The subscript r is not the index of the variable.)

4. Calculate the flow rate in each ring (corresponding to equation 5.7) by entering D<Ctrl+g>Q.r[0,j:2*p<Ctrl+g>*r[0,j *(r[0,j+1 -r[0,j)*v.r[0,j. (Note that three spaces must be inserted after "j+1" and two spaces must be inserted between "j" and ")" in this expression.) The key sequence D<Ctrl+g>Q.r creates the variable ΔQ_r, and the sequence [0,j identifies it as the flow rate corresponding to the *j*th location. Of the terms on the right side of the equation, r[0,j represents the radial location r, (r[0,j+1 -r[0,j) represents the thickness of the ring, and v.r[0,j represents the velocity corresponding to that radial location. As seen from Figure 5.14, ΔQ_r is a matrix containing 1 row and 10 columns, with the numbers representing the flow in each differential element. The range of index j needs to be defined before these flow rate calculations can be performed; type j:0;9. To obtain the total flow rate Q by summation of the ΔQ_r values, type Q:<Ctrl+4>D<Ctrl+g>Q.r. The <Ctrl+4> sequence inserts the summation operator in the Mathcad worksheet.
5. Calculate the average velocity and the Reynolds number by entering the appropriate formulae. As mentioned previously, the units by default are SI units and need to be changed if the answers are desired in other units.

It can be seen that the solution using Mathcad differs slightly from that obtained using Excel. The explanation for this discrepancy is left to the reader.

Example 5.4 Shear Force on Pipe Wall

For the flow problem in Example 5.3, what is the shear force exerted at the pipe surface? Assume the pipe length to be 2 m.

Solution algorithm

The shear force, F_{shear}, is obtained by multiplying the shear stress by the area of the surface (A_{shear}) over which it acts. From equation 5.2, we get the following:

$$F_{shear} = \mu\left|\left(\frac{dv_r}{dr}\right)\right| \cdot A_{shear} \tag{E5.7}$$

Here A_{shear} is given by the following:

$$A_{shear} = 2\pi RL \tag{E5.8}$$

(*Continues*)

Example 5.4 Shear Force on Pipe Wall (*Continued*)

The velocity gradient is obtained from the data in Table 5.2. The following is used for numerical differentiation:

$$\left|\left(\frac{dv_r}{dr}\right)\right| = \left|\frac{v_{r=R} - v_{r=0.9R}}{R - 0.9R}\right| \tag{E5.9}$$

Solution (using a spreadsheet program)

The calculations using Excel are shown in Figure 5.16.

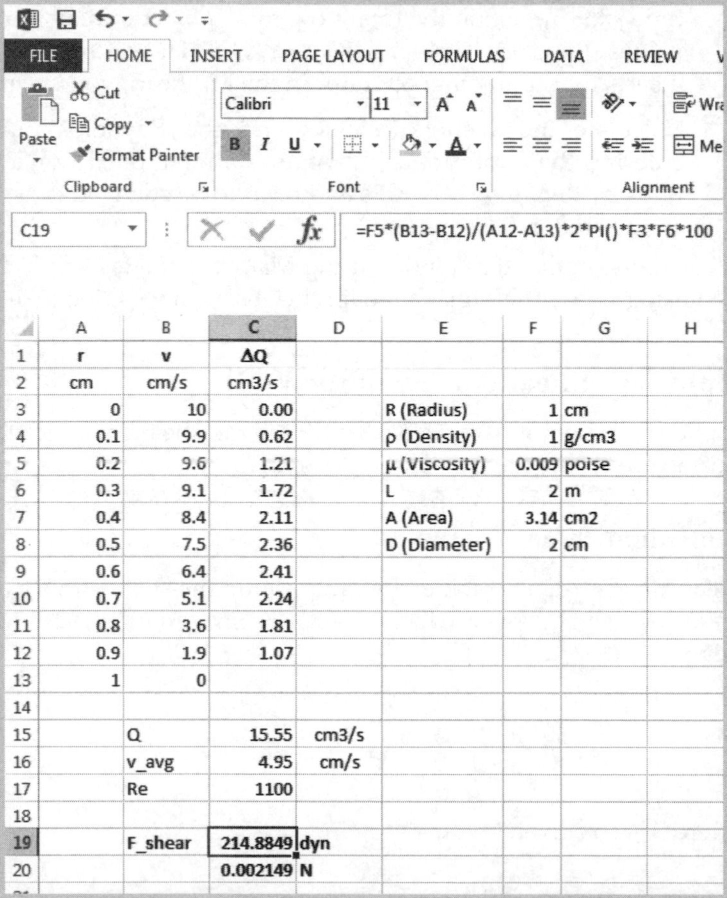

FILE HOME INSERT PAGE LAYOUT FORMULAS DATA REVIEW

Cut Copy Format Painter Paste | Calibri 11 | Clipboard Font Alignment

C19 =F5*(B13-B12)/(A12-A13)*2*PI()*F3*F6*100

	A	B	C	D	E	F	G	H
1	r	v	ΔQ					
2	cm	cm/s	cm3/s					
3	0	10	0.00		R (Radius)	1	cm	
4	0.1	9.9	0.62		ρ (Density)	1	g/cm3	
5	0.2	9.6	1.21		μ (Viscosity)	0.009	poise	
6	0.3	9.1	1.72		L	2	m	
7	0.4	8.4	2.11		A (Area)	3.14	cm2	
8	0.5	7.5	2.36		D (Diameter)	2	cm	
9	0.6	6.4	2.41					
10	0.7	5.1	2.24					
11	0.8	3.6	1.81					
12	0.9	1.9	1.07					
13	1	0						
14								
15		Q	15.55	cm3/s				
16		v_avg	4.95	cm/s				
17		Re	1100					
18								
19		F_shear	214.8849	dyn				
20			0.002149	N				

Figure 5.16 Shear force calculation using Excel.

The length of pipe is shown in cell F6. The shear force calculation is in cell C19 with the formula as shown. The conversion factor of 100 is used to convert the length units from meter to centimeter. The shear force obtained is ~214 dyn. The answer unit is converted into SI unit force (newtons, or N) using a conversion factor of 1 dyn = 10^{-5} N.

Solution (using Mathcad)

The solution involves inserting the following statements in the previous Mathcad document:

1. L:2*m
2. F.shear:m<Ctrl+g>*(v.r[0,10 -v.r[0,9)/r[0,9 -r[0,10 *2*p<Ctrl+g>*R*L

   (Note that two spaces are inserted before "-v.r"; one space before ")"; one space before "-r"; and four spaces before "*2*".)

The shear force is calculated to be 2.149×10^{-3} N, as seen in Figure 5.17.

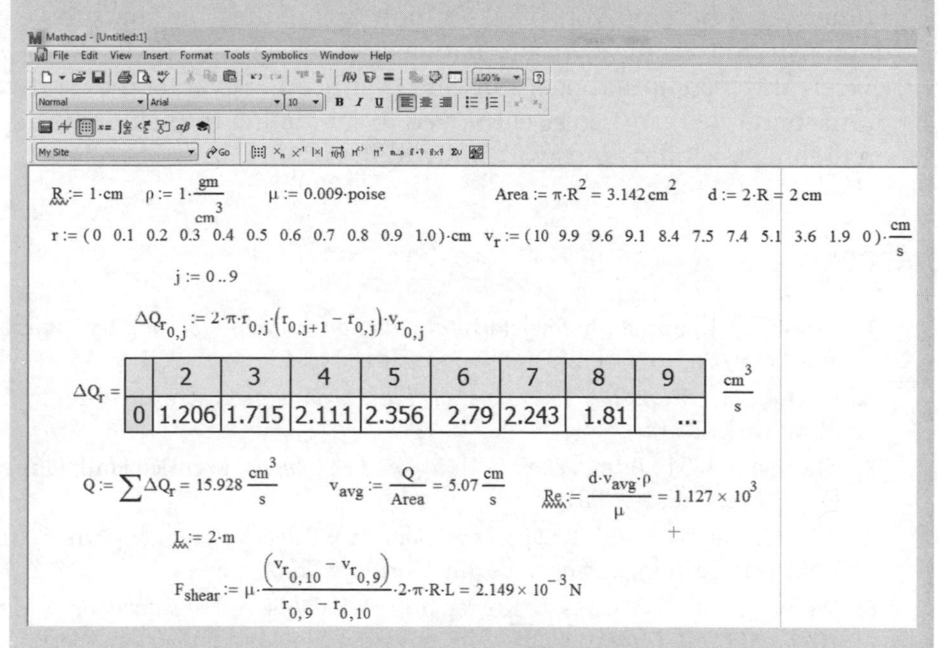

Figure 5.17 Shear force calculation using Mathcad.

The result is identical to that obtained using Excel. Entering the formula in Excel is slightly easier than in Mathcad. On the other hand, Mathcad offers automatic conversion of units, removing the need for manually entering the conversion factors.

The accuracy of derivative computations increases with a decrease in the thickness of the ring. Additional radial position-velocity data availability at intervals of 0.05 cm rather than 0.1 cm would increase the number of computations but enable us to estimate the velocity derivative more accurately. The flow rate computations would also be more accurate.

5.4 Summary

This chapter presented the elementary principles of fluid flow both qualitatively and quantitatively. Examples of various computation problems were presented. The problems included solution of transcendental equations, regression analysis and interpolation, numerical integration, and numerical differentiation. The solution techniques for the problems were demonstrated using two alternative approaches: a spreadsheet program (Excel) and a computational software (Mathcad). Several other software programs and tools can also be employed for obtaining the solutions, depending on their accessibility and availability to the user.

References

1. Olson, A. T., and K. A. Shelstad, *Introduction to Fluid Flow and the Transfer of Heat and Mass*, Prentice Hall, Englewood Cliffs, New Jersey, 1987.
2. Oertel, H., *Prandtl's Essentials of Fluid Mechanics*, Third Edition, Springer, New York, 2010.
3. Thomson, W. J., *Introduction to Transport Phenomena*, Prentice Hall, Upper Saddle River, New Jersey, 2001.
4. Schlichting, H., and K. Gersten, *Boundary Layer Theory*, Eighth Revised and Enlarged Edition, Springer, Berlin, Germany, 2000.
5. Welty, J., C. E. Wicks, G. L. Rorrer, and R. E. Wilson, *Fundamentals of Momentum, Heat and Mass Transfer*, Fifth Edition, John Wiley and Sons, New York, 2008.

Problems

5.1 Calculate the Reynolds numbers for a 1.5 in. inside diameter pipe carrying water at a flow rate of 0 to 5 gpm. Assume a temperature of 25°C.

5.2 Calculate the Reynolds numbers for the following situation: (a) a 1 μm sized microbe swimming with a speed of 30 μm/s; (b) a swimmer competing in an Olympic 100 m race finishing in 50 s. Make any reasonable assumptions necessary for the solution.

5.3 The viscosity of 30 wt engine oil at 100°C is 0.0924 poise. What is the viscous (shear) force needed to slide an 8 cm diameter, 8 cm long piston through a cylinder on a 2 μm thick oil film with a speed of 8 m/s?

5.4 For noncircular geometries, a hydraulic diameter (D_h) is used as the characteristic length parameter for calculating the Reynolds number calculation:

$$D_h = \frac{4 \cdot \text{Cross sectional Area}}{\text{Wetted Perimeter}}$$

An HVAC duct circulates 600 cfm (cubic feet per minute) of air at 85°F through an 18 in. × 12 in. rectangular duct. What is the air velocity? What is the Reynolds number if the air density and viscosity at 85°F are 1.177 kg/m^3 and 1.85×10^{-2} mPa·s, respectively?

5.5 It may be feasible to extract uranium from sea water (concentration 8 ppb, or parts per billion) by placing uranium absorbing plates in the ocean. The oceanic waves result in circulation of water through the absorbing structure. The absorbing plates are mounted so they divide a 5 ft × 5 ft square pipe into smaller square pipes. The volumetric flow rate through the big pipe is 1000 gal/min. What is the Reynolds number if the viscosity is 1 centipoise (cP)? How small can the opening of the smaller square be for the flow to still be in the turbulent regime? Assume that the average velocity remains constant.

5.6 Supertankers with ultra-large capacity can carry 3,166,353 barrels of crude oil (1 barrel = 42 gal). If it takes two days to unload the tanker using a 24 in. diameter hose, what is the average velocity of the oil? If the oil density and viscosity are 870 g/L and 0.043 poise, respectively, what is the Reynolds number? Is this flow turbulent?

5.7 What is the friction factor for the flow in problem 5.6? What is the pressure drop if the oil is pumped into a storage tank located 1 km away?

5.8 The height h of a liquid in a partially filled spherical tank (diameter D) can be calculated from the following equation:

$$V_{liquid} = \frac{\pi}{3} h^2 (1.5D - h)$$

What is the height of liquid when a 35,000 m^3 capacity storage tank is only 75% full? What is the area of the free liquid surface?

5.9 The following data were obtained for the pressure drop for water flow through 100 ft of fire hose:

Hose Diameter, in.	Flow Rate, gpm	Pressure Drop, psi
1.5	120	45
2.0	150	20
2.5	220	12
3.0	400	15

Plot the pressure drop as a function of the Reynolds number for the flow. Make reasonable assumptions in order to perform these calculations. Calculate the friction factors from these data and compare those with the friction factors obtained using the Nikuradse equation.

5.10 Enbridge Line 5 is a 30 in. pipeline connecting Wisconsin and Canada through Michigan's upper and lower peninsulas carrying ~550,000 barrels of light crude every day. The line is split into two 20 in. pipes buried under deep water for crossing 4.5 miles of the Straits of Mackinac. Assuming the fluid densities and viscosities of problem 5.6, calculate the Reynolds numbers for both the 30 in. and 20 in. sections. What is the pressure drop across the Straits of Mackinac?

CHAPTER 6

Material Balance Computations

In nature there is no annihilation;
And therefore the thing which is consumed
Either passes into air, or
Is received into some adjacent body.

—Sir Francis Bacon[1]

The fundamental principle of the *conservation of mass* states that matter can neither be created nor destroyed—a concept that has been introduced to students sometimes as early as elementary school. As matter cannot be destroyed, any material or part of material removed from any location must appear at some other location, as articulated so eloquently by Sir Francis Bacon. Accounting for the material, down to its last molecule (or atom), is absolutely critical for an industrial process, since there is a monetary value associated with the chemical species that constitute the matter. Further, accounting of chemicals is also needed because a large number of chemicals are hazardous in nature and have potential to cause harm to the environment, human health, and ecosystems. Proper handling and management of process streams requires a knowledge of the chemicals present in the streams and their quantities and compositions. Balancing the inflow of the matter with the outflow is the foundation of chemical engineering, and a chemical engineer is expected to master the art of *material balance* [1]. This chapter first presents the general principles of material balance, then introduces some of the typical material balance computations.

6.1 Quantitative Principles of Material Balance

A chemical process typically consists of several units that may involve chemical reactions and/or simple physical separation and mixing operations, as described in previous chapters. The process streams may be constituted of a single phase (gas/liquid/solid) or may be multiphase in nature. A unit may

1. Sixteenth-century English philosopher and scientist, widely regarded as the Father of the Scientific Method. Quotation source: Spedding, J., R. L. Ellis, and D. D. Heath, *The Works of Francis Bacon*, Vol. 2: Philosophical Works 2, Cambridge University Press, Cambridge, England, 1857.

or may not be operating at steady state. Regardless of the situation, the units and the process are amenable to the material balance analysis based on the common principles described in the sections that follow.

6.1.1 Overall Material Balance

Consider an arbitrary process represented by the block flow diagram shown in Figure 6.1. The process unit has two influent flows (streams 1 and 2) feeding into a tank and one effluent flow (stream 3) leaving the tank.

Physically, the material being fed to the process must either accumulate in it or exit the system. The principle of conservation of mass dictates that the total mass fed to the system must equal the sum of the mass exiting the system *and* the mass accumulating in the system [2, 3]:

Rate of Mass Input to the Process = Rate of Mass Accumulation in the Process + Rate of Mass Output from the Process

If $\dot{m}_1$, $\dot{m}_2$, and $\dot{m}_3$ are the mass flow rates of the three streams, and m_S is the total mass in the process unit,[2] then

$$\dot{m}_1 + \dot{m}_2 = \frac{dm_S}{dt} + \dot{m}_3 \tag{6.1}$$

Equation 6.1 is the mathematical representation of the *overall material balance* for the process. If the rate at which material is taken out of the system, $\dot{m}_3$, is smaller than the rate at which material is being fed to the system, $\dot{m}_1 + \dot{m}_2$, material will accumulate in the process, increasing the system mass m_S, as would be the case during process start-up. On the other hand, if $\dot{m}_3$ is greater than $\dot{m}_1 + \dot{m}_2$—that is, material is removed at a faster rate from the process

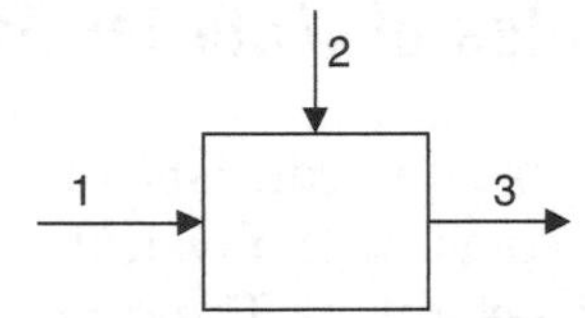

Figure 6.1 A simple process unit.

2. By convention, a variable with a dot placed on top indicates a rate, so while m represents the mass (g, kg, and so on), $\dot{m}$ represents the mass rate (g/s, kg/h, and so on).

than being fed—then m_S will decrease with time, as in the case of draining a tank. Generalizing for multiple input and output streams, the overall material balance equation is as follows:

$$\sum_i \dot{m}_{inlet,i} = \sum_j \dot{m}_{outlet,j} + \frac{dm_S}{dt} \tag{6.2}$$

Here, $\dot{m}_{inlet,i}$ and $\dot{m}_{outlet,j}$ represent the mass flow rates of the *i*th inlet and *j*th outlet streams. As mentioned previously, a large number of chemical processes operate at steady state, meaning that the conditions are invariant with respect to time. The overall material balance for such steady-state processes is then simplified to equation 6.3.

$$\sum_i \dot{m}_{inlet,i} = \sum_j \dot{m}_{outlet,j} \tag{6.3}$$

The overall balance clearly serves a valuable purpose in material accounting. The discrepancy between mass inflow and outflow may be used to estimate atmospheric fugitive emissions and leakages and to identify malfunctioning process equipment.

6.1.2 Component Material Balance

Chemical engineers require additional information (apart from the overall material balance) about material flows of components in the process and conduct *component material balances* over the process units. Let us assume that the process shown in Figure 6.1 is that of simple mixing of a concentrated aqueous solution of salt *A* (stream 1) with pure water (H_2O; stream 2), yielding a dilute aqueous solution of the salt (stream 3). This is a two-component system composed of the components *A* and H_2O. Applying the principle of the conservation of mass to each component leads to the two component balances shown in equations 6.4 and 6.5.

$$\dot{m}_{A,1} = \dot{m}_{A,3} + \frac{dm_A}{dt} \tag{6.4}$$

$$\dot{m}_{H_2O,1} + \dot{m}_{H_2O,2} = \dot{m}_{H_2O,3} + \frac{dm_{H_2O}}{dt} \tag{6.5}$$

Here, m_A and m_{H_2O} represent the masses of component *A* and H_2O present in the system. Since component *A* is present in only one inlet stream, the left side of equation 6.4 involves only one inlet term. H_2O, however, is present in both of the inlet streams, and hence, the left side of equation 6.5 has two terms. The generalized component balance equations for an n component, multistream unit can be written as follows:

$$\sum_{i} \dot{m}_{k,i} = \sum_{j} \dot{m}_{k,j} + \frac{dm_k}{dt}; k = 1,2,..,\text{n} \tag{6.6}$$

The left side of this equation represents the mass of component k being fed to the unit through all the inlet streams (summation over i inlet streams), the first term on the right side represents the mass flow rate of component k out of the system through all outlet streams (summation over j outlet streams), and the last term represents the rate of accumulation of the component mass in the system. The accumulation term will drop out of the equation for a continuous, steady-state process, simplifying the equation from a first-order ordinary differential equation to an algebraic or a transcendental equation.

In total there will be n equations representing component balances for the n components. Thus, in an n-component system, we have n independent component balance equations and one overall material balance equation—a total of $n + 1$ equations. However, only n of these equations are independent (distinct), as the summation of all component balances will lead to the overall material balance—equation 6.2.

The application of these principles for solving material balance problems for systems that involve only physical operations is described in section 6.2, followed by the application to reacting systems in section 6.3.

6.2 Material Balances in Nonreacting Systems

Material balances in nonreacting systems are illustrated through a simple example involving dilution of a concentrated solution, an operation frequently encountered in a chemical process plant.

Example 6.2.1 Dilution of a Concentrated Aqueous Solution

The production process for sodium hydroxide (NaOH) yields a 28% (by mass) solution of sodium hydroxide in a membrane cell. A subsequent process requires 1000 tons per day (tpd) of 10% NaOH solution. Calculate the quantities of the concentrated solution and diluent H_2O needed to obtain this stream.

Solution

The process can be accurately represented by Figure 6.1. Let us assume that the process operates continuously at steady state, and $\dot{m}_1$, $\dot{m}_2$, and $\dot{m}_3$ are the mass flow rates of the concentrated (28%) solution, diluent H_2O, and dilute solution, respectively. This is a system containing two components—NaOH and H_2O—and the resultant overall and component material balances are shown in equations E6.1, E6.2, and E6.3.

$$\text{Overall balance:} \quad \dot{m}_1 + \dot{m}_2 = 1000 \quad \text{(E6.1)}$$

$$\text{NaOH balance:} \quad 0.28 \cdot \dot{m}_1 = 0.1 \cdot 1000 \quad \text{(E6.2)}$$

$$H_2O \text{ balance:} \quad 0.72 \cdot \dot{m}_1 + \dot{m}_2 = 0.9 \cdot 1000 \quad \text{(E6.3)}$$

Of these three equations, only two are independent and sufficient for quantitative solution in this two-component system. The two equations to be used are the overall material balance and the component material balance for NaOH. It is preferable to use the component balance for NaOH rather than H_2O, as it is present in fewer streams than is H_2O. The solution steps are

1. Equation E6.2 is solved first to obtain the value of $\dot{m}_1$, which is 357.1 tpd.
2. This value is substituted in equation E6.1 to obtain 642.9 tpd as the value of $\dot{m}_2$.

Therefore, 642.9 tpd of H_2O needs to be added to 357.1 tpd of 28% NaOH solution to obtain 1000 tpd of the desired 10% NaOH solution. The validity of this solution can be (and should be) cross-checked with the component balance for H_2O. The left side of equation E6.3 evaluated from the calculated values of $\dot{m}_2$ and $\dot{m}_3$ is found to be 900 tpd, equal to the right side of the equation.

Dilution of a concentrated solution is a type of mixing operation, a common operation in chemical processes. Equally common are separation operations, which can be visualized as the inverse of mixing operations. In the simplest of separations, a single stream containing two components is separated into two streams, each with a different composition, as illustrated in example 6.2.2.

Example 6.2.2 Separation by Distillation

Distillation involves separating two or more components based on their volatilities. A 1000 kg/h stream containing 50% benzene and 50% toluene (by mass) is fed to a distillation column to obtain a benzene-rich stream at the top (distillate) and a toluene-rich stream at the bottom (bottoms). Ninety percent of the benzene fed to the column is recovered in the distillate, which has a purity of 99%. Calculate the mass and molar flow rates and compositions of both the product streams.

Solution

The schematic of this operation is shown in Figure 6.2. F, D, and B are the flow rates of the feed, distillate, and bottoms, respectively. The compositions (mass fractions) in the three streams are represented by z, y, and x, with the subscripts B and T representing benzene and toluene, respectively. The overall and component balances are shown in equations E6.4, E6.5, and E6.6.

Overall balance: $$B + D = F = 1000 \tag{E6.4}$$

Benzene balance: $$F \cdot z_B = B \cdot x_B + D \cdot y_B \tag{E6.5}$$

Toluene balance: $$F \cdot z_T = B \cdot x_T + D \cdot y_T \tag{E6.6}$$

As mentioned earlier, only two of these equations are independent, with the toluene balance obtained by simply subtracting the benzene balance from the overall balance. The benzene balance can be further simplified using the information about the purity of the distillate stream. Substituting values of F, z_B, and y_B, in equation E6.5 reduces to the following:

$$500 = B \cdot x_B + D \cdot 0.99 \tag{E6.7}$$

To solve the system of these equations, the additional information related to the recovery of benzene must be used. Since 90% of benzene fed to the column is recovered in the distillate, the following applies:

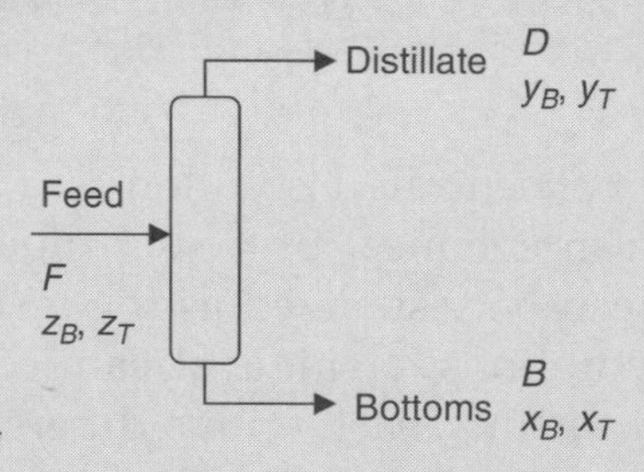

Figure 6.2 Distillation of benzene-toluene mixture.

$$500 \cdot 0.9 = D \cdot 0.99 \quad \text{(E6.8)}$$

Equation E6.8 yields the value of D to be 454.5 kg/h, which upon substitution in equation E6.4 yields B as 545.5 kg/h. The mass fraction of benzene in the bottoms, x_B is then obtained from equation E6.7 to be 0.09. The mass fractions of toluene in the distillate and bottoms streams are obtained by subtracting mole fractions of benzene from 1, resulting in $x_T = 0.91$ and $y_T = 0.01$. The accuracy of these numbers is checked by evaluating the right side of equation E6.6:

$$B \cdot x_T + D \cdot y_T = 545.5 \times 0.91 + 454.5 \times 0.01 = 500.95 \text{ kg/h } (\sim 500 \text{ kg/h})$$

This calculation indicates that the values obtained by solving equations E6.4 and E6.5 satisfy equation E6.6, confirming the validity of the solution. The molar flow rates of the components can be obtained by dividing their mass flow rates by the molar masses (78 g/mol for benzene, 92 g/mol for toluene). Mole fractions of components are then obtained by dividing the molar flow rate of the component by the total molar flow rate. The resultant flow rates and compositions of various streams are shown in Tables E6.1 and E6.2, respectively.

Table E6.1 Flow Rates of Streams

	Feed, *F*		**Distillate, *D***		**Bottoms, *B***	
	kg/h	**mol/h**	**kg/h**	**mol/h**	**kg/h**	**mol/h**
Benzene	500	6410	450	5769	50	641
Toluene	500	5435	4.5	49	495.5	5386
Total	1000	11845	454.5	5827	545.5	6027

Table E6.2 Fractional Stream Compositions

	Feed, *z*		**Distillate, *y***		**Bottoms, *x***	
	mass	**mole**	**mass**	**mole**	**mass**	**mole**
Benzene	0.5	0.54	0.99	0.99	0.09	0.11
Toluene	0.5	0.46	0.01	0.01	0.91	0.89
Total	1.0	1.0	1.0	1.0	1.0	1.0

These calculations can be easily performed on a basic calculator or using an appropriate software package. The Mathcad solution is shown in Figure 6.3.

(Continues)

Example 6.2.2 Separation by Distillation (*Continued*)

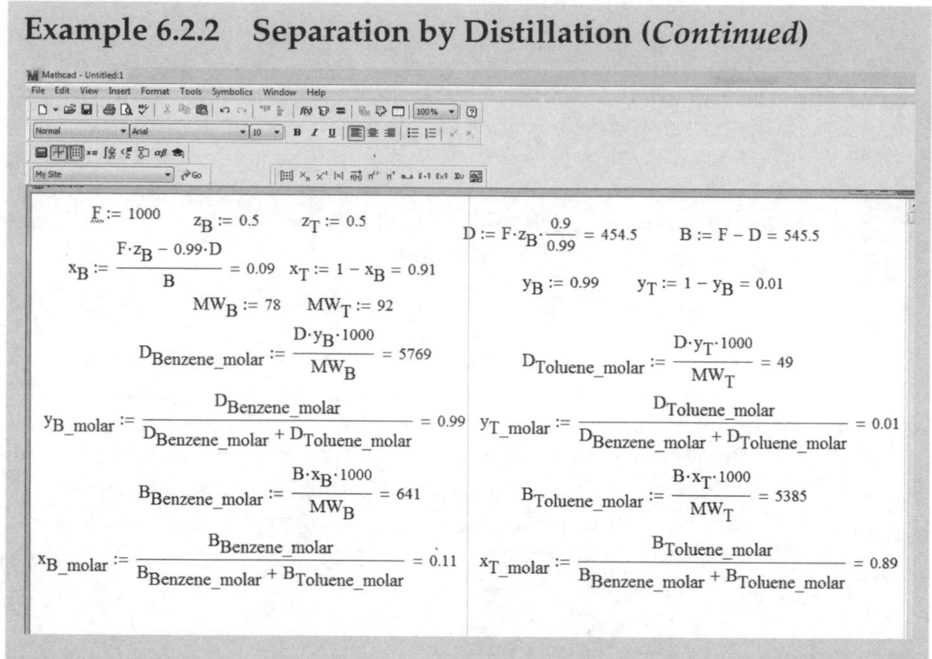

Figure 6.3 Mathcad solution of problem 6.2.2.

The benefit of using Mathcad lies in the ability to obtain the solution rapidly when the problem specifications—stream purity, component recovery, and so on—are changed. Any change in the variable is immediately reflected in the solution.

These two examples involve simple physical processes of mixing and separation. The principles and procedures used therein can readily be extended to multicomponent, multistream processes. The computations for solutions in these examples are illustrated using a *mass* basis; that is, quantities of various species are expressed in terms of mass units. However, for a nonreacting system, the computations can be performed equally well on a *molar* basis; that is, by expressing quantities of various species in terms of the number of moles present. The nonreacting systems are amenable to computations on a molar basis, as no new chemical species are created, nor any existing ones destroyed. In fact, computations using molar basis are much more common than those using a mass basis, as most relationships between chemical species are expressed typically in terms of molar concentrations. It is, of course, possible to convert these relationships on a mass basis; however, that introduces a level of complexity that is neither necessary nor desired.

Not all chemical species are conserved in a reacting system, which also involves creation of new species. The application of the overall and component material balances to chemically reacting systems involves a slightly higher level of complexity than that for a nonreacting system, as explained in the next section.

6.3 Material Balances in Reacting Systems

Reacting systems are characterized by the disappearance of reactants and appearance of the products. Clearly, molecular species are not conserved in the reaction, and at steady state, the rate of input of a compound participating in the reaction is not equal to the rate of its output. However, *as long as no nuclear reactions are occurring in the system*, atomic species are conserved. Therefore, the approach to obtaining the material balance in a reacting system involves formulating independent conservation equations for the atoms of elemental species involved in the reactions [4]. It should also be noted that several inert species that do not participate in the reaction are often present in a reacting system, and the material balance for such species can be expressed in terms of molecular quantities. For example, if a combustion reaction is conducted using air as the oxidant, then the nitrogen (N_2) present in air does not participate in the reaction and is treated as an inert. Similarly, solvents employed for liquid-phase reactions—water for aqueous phase reactions, for example—do not participate in the reaction, and the material balance for these inert solvents may be written in terms of molecular species.

Conducting a material balance on a reacting system requires a knowledge of the balanced chemical equation(s) for the reaction(s). The chemical equation provides the information about the proportions of the different species involved in the reaction—the stoichiometry of the reaction. Let us consider a general reaction involving species A, B, C, and D according to the equation:

$$a\mathrm{A} + b\mathrm{B} \rightarrow c\mathrm{C} + d\mathrm{D} \tag{6.7}$$

where a, b, c and d are the stoichiometric coefficients of the species A, B, C, and D, respectively. According to this equation, the ratio of the amount of C formed (molecules or moles) to the amount of A reacting will be equal to c/a. Similarly, proportional quantities of D are formed and B are reacted, depending on the stoichiometric coefficients d and b, respectively. Example 6.3.1 illustrates the application of these principles to a simple reacting system involving a single reaction.

Example 6.3.1 Combustion of Natural Gas

Natural gas is burned in a reactor (such as in a typical gas water heater in a home) using air as the oxidant. Calculate the quantities of air needed and the products formed per mole of the natural gas burned in the system.

Solution

Obtaining the material balance for this system requires knowledge of the chemicals present in the natural gas. Natural gas is, in reality, a mixture of gases dominated by methane (CH_4). However, in these calculations, it is assumed that the natural gas is composed entirely of CH_4. This assumption allows us to simplify the system and write a single chemical reaction to represent the combustion. Complete combustion of CH_4 results in the formation of carbon dioxide (CO_2) and H_2O according to the following equation:

$$CH_4 + 2O_2 \rightarrow CO_2 + 2H_2O$$

From the equation, the stoichiometric proportions between various species are obtained as follows:

$$O_2/CH_4 = 2,\ CO_2/CH_4 = 1,\ \text{and}\ H_2O/CH_4 = 2$$

Therefore, two moles of oxygen (O_2) are needed per mole of natural gas, and the 3 moles of product are formed that include 1 mole of CO_2 and 2 moles of H_2O. Since the O_2 supplied is obtained from air, the system also contains N_2.[3] The amount of N_2 associated with O_2 is equal to 2 (mol O_2) · 79/21 (mole ratio of N_2 to O_2 in air), that is, 7.52 moles per mole of natural gas. All of this N_2 passes unreacted through the system and exits with the product gas. The complete material balance for the system is shown in Figure 6.4.

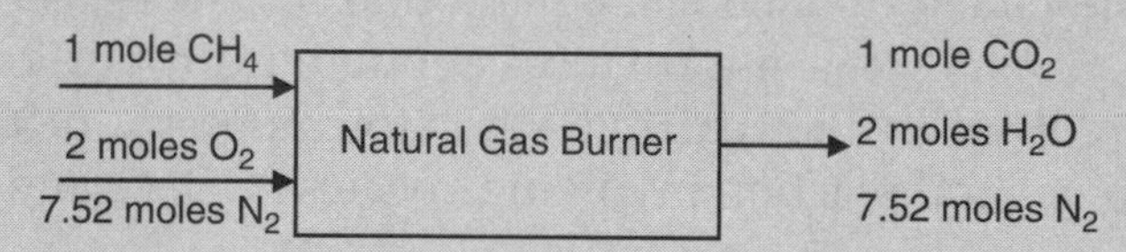

Figure 6.4 Material balance on a natural gas burner.

---

3. This is a simplifying assumption regarding the composition of air. Obviously, air has many more constituents in addition to O_2 and N_2. However, unless required by rigorous specifications, the trace constituents are neglected and air is considered as a mixture of N_2 and O_2 with the molar ratio of N_2 to O_2 being 79/21 (= 3.76).

It can be seen that all the CH_4 and O_2 fed to the burner react completely and are not present in the outlet product stream. This is an idealized representation, because an actual operating burner may have some inefficiencies that result in incomplete combustion of CH_4—which will result in some unreacted CH_4 being present in the product stream, or more likely, some fraction of the CH_4 fed is converted to carbon monoxide (CO) rather than CO_2. In any case, there will be unreacted O_2 present in the product stream. Further, excess O_2 is typically provided in combustion reactions to accomplish a complete conversion of the fuel. These situations will result in a material balance modified as follows:

Modification 1 Excess Air Supplied

Rework the material balance when the air supplied is 15% in excess of the stoichiometric requirements.

Solution

The stoichiometric O_2 requirement is as given by the previous equation:

Stoichiometric O_2 requirement: 2 moles/mole of CH_4

Total O_2 supplied: 2.30 (= 2 × 1.15) moles/mole of CH_4

Total N_2 fed with O_2: 8.65 (= 2.30 × 79/21) moles/mole CH_4

This N_2 will pass unreacted through the system, ending up in the product stream along with 1 mole of CO_2 and 2 moles of H_2O formed from CH_4. In addition, the product stream will contain unreacted O_2, the quantity of which is 0.30 moles (2.30 moles fed—2 moles reacted according to the equation). The resulting material balance is shown in Figure 6.5.

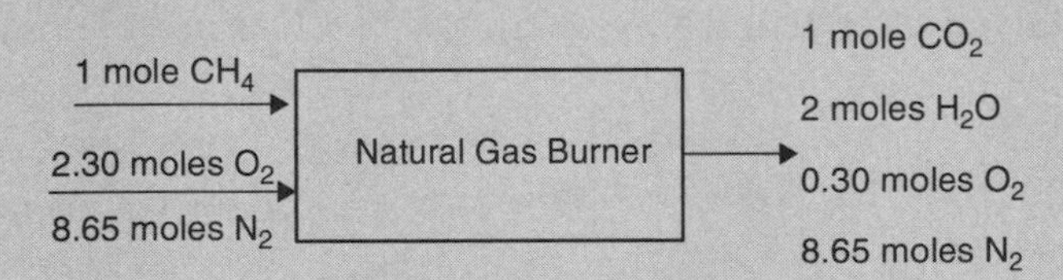

Figure 6.5 Material balance on a natural gas burner—excess air.

Modification 2 Excess Air Supplied and Incomplete Combustion of Methane

Rework the material balance when 1% of the CH_4 fed does not react in the burner. Further, of the CH_4 burned, only 90% undergoes complete combustion. Fifteen percent excess air is supplied.

Solution

The product gas mixture will now contain a total of six components: unreacted CH_4, CO_2, CO, H_2O, N_2, and O_2. The quantity of air fed is that calculated earlier for modification 1. To calculate the quantities of various species formed and O_2 reacted and remaining, a balanced chemical equation for incomplete combustion of CH_4 (formation of CO, rather than CO_2) is needed:

$$CH_4 + 1.5O_2 \rightarrow CO + 2H_2O$$

From the given information, 1% of CH_4 does not burn, so CH_4 in the product stream is equal to 0.01 moles; 0.99 moles of CH_4 react, 90% according to the equation corresponding to complete combustion and 10% according to that corresponding to incomplete combustion to CO. Therefore,

CO_2 formed: $0.99 \times 0.90 = 0.891$ moles

CO formed: $0.99 \times 0.10 = 0.099$ moles

H_2O formed: $0.99 \times 0.90 \times 2 + 0.99 \times 0.10 \times 2 = 1.98$ moles

O_2 reacted: $0.99 \times 0.90 \times 2 + 0.99 \times 0.10 \times 3/2 = 1.9305$ moles

Hence, unreacted $O_2 = 2.30 - 1.9305 = 0.3695$ moles. Figure 6.6 shows the resulting material balance.

This solution was obtained through stoichiometric proportions given by the two equations. Atomic species balances are implicit in these proportions. The same solution can be obtained through explicit atomic species balances. For example, the amounts of O_2 and N_2 in the feed are calculated as before. The amounts of various species in the outlet stream are unknown at this point and are denoted by unknown variables: n_1 (moles of CH_4), n_2

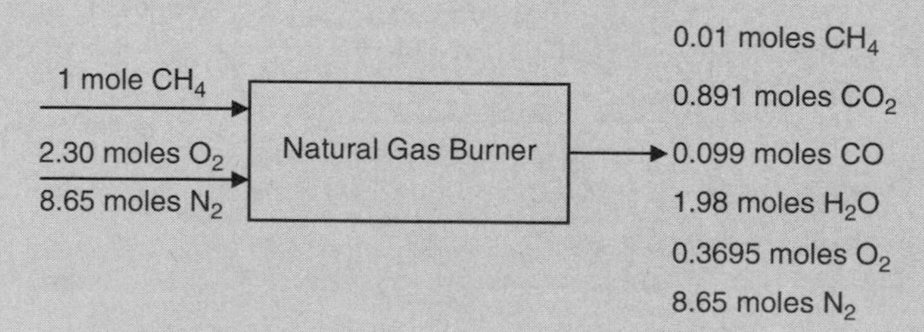

Figure 6.6 Material balance on a natural gas burner—excess air and incomplete combustion.

(moles of CO_2), n_3 (moles of CO), n_4 (moles of H_2O), n_5 (moles of O_2), and n_6 (moles of N_2). Balance on each of the atomic species (C, H, O, and N) yield the following four equations:

$$\text{C balance:} \quad n_1 + n_2 + n_3 = 1 \tag{E6.9}$$

$$\text{H balance:} \quad 4n_1 + 2n_4 = 4 \tag{E6.10}$$

$$\text{O balance:} \quad 2n_2 + n_3 + n_4 + 2n_5 = 4.6 \tag{E6.11}$$

$$\text{N balance:} \quad 2n_6 = 17.3 \tag{E6.12}$$

Equation E6.12 yields the moles of N_2 in the outlet, n_6, directly as 8.65, leaving three equations (E6.9, E6.10, and E6.11) in five unknowns. This system of equations cannot be solved unless more information provided in the problem statement is used to formulate two additional equations.[4] The first equation is obtained from the information that only 99% of the CH_4 is reacted. Since 1% of CH_4 has not reacted, the following applies:

$$n_1 = 0.01 \tag{E6.13}$$

The second equation is obtained from the information that 90% of the CH_4 reacted undergoes complete combustion (it is converted to CO_2), and the remaining 10% undergoes incomplete combustion (it is converted to CO). Therefore, the molar ratio between CO_2 to CO must be 9:

$$n_2/n_3 = 9 \tag{E6.14}$$

Equations E6.9, E6.10, E6.11, E6.13, and E6.14 form a system of five equations in five unknowns that are rewritten in matrix form as follows (equation E6.14 is reformatted as $n_2 - 9n_3 = 0$ before entering it into the matrix form):

$$\begin{bmatrix} 1 & 0 & 0 & 0 & 0 \\ 1 & 1 & 1 & 0 & 0 \\ 0 & 1 & -9 & 0 & 0 \\ 4 & 0 & 0 & 2 & 0 \\ 0 & 2 & 1 & 1 & 2 \end{bmatrix} \begin{bmatrix} n_1 \\ n_2 \\ n_3 \\ n_4 \\ n_5 \end{bmatrix} = \begin{bmatrix} 0.01 \\ 1 \\ 0 \\ 4 \\ 4.6 \end{bmatrix} \tag{E6.15}$$

(*Continues*)

4. The *degrees of freedom analysis*, typically covered in the material and energy balances course, discusses in detail the relationship between unknown variables, mass balance equations, and other equations and specifications needed for complete material and energy balance solution for the system. Here, we simply try to formulate as many equations as there are unknown variables.

Modification 2 Excess Air Supplied and Incomplete Combustion of Methane (*Continued*)

Equation E6.15 conforms to the matrix form **[A][X]** = **[B]** for simultaneous linear equations, where **[A]** is the matrix of coefficients, **[X]** is the vector of unknown variables, and **[B]** is the vector matrix of values (those on the right) of the equations. Several algorithms are available for obtaining solutions for such systems of equations, as described in Chapter 4. One of the common algorithms involves taking the inverse of the matrix of coefficients, and multiplying it by matrix **[B]**:

$$[X] = [A]^{-1}[B] \tag{E6.16}$$

Several software programs allow execution of mathematical operations involving matrices, such as computation of a matrix inverse and matrix multiplication. The solution techniques using a spreadsheet program and Mathcad follow.

Solution (using a spreadsheet program)

The solution using Excel is shown in Figure 6.7. The first step in the solution is entering the matrix of coefficients **[A]** and the vector matrix of values **[B]**. As seen from Figure 6.7, **[A]** forms a 5 × 5 array occupying cells B5 through F9, while **[B]** is entered in cells H5 through H9. To calculate $[\mathbf{A}]^{-1}$, first a 5 × 5 space is selected—B12 through F16, as seen previously—and the matrix inverse function is entered by typing =MINVERSE(B5..F9) followed by pressing the keys <Ctrl+Shift+Enter> simultaneously. Excel calculates the inverse of the matrix and populates the cells B12 through F16 automatically. The solution to the material balance problem is then obtained by multiplying the matrices $[\mathbf{A}]^{-1}$ and **[B]**. This is accomplished by selecting a 5 × 1 space (H12–H16), entering the formula =MMULT(B12..B16,H5..H9), and pressing the keys <Ctrl+Shift+Enter> simultaneously. The resultant solution of the problem is seen in the cells H12 through H16, which are the values for the number of moles of CH_4, CO_2, CO, H_2O, and O_2—0.01, 0.891, 0.099, 1.98, and 0.3695, respectively.

Solution (using Mathcad)

The system of linear equations **[A][X]** = **[B]** can be readily solved in Mathcad using the function lsolve(**[A]**,**[B]**).[5] Figure 6.8 shows the Mathcad solution of the problem.

5. The algorithm used in the Mathcad *lsolve* function is called *LU decomposition*, a type of elimination algorithm mentioned in Chapter 4, the details of which are beyond the scope of this book.

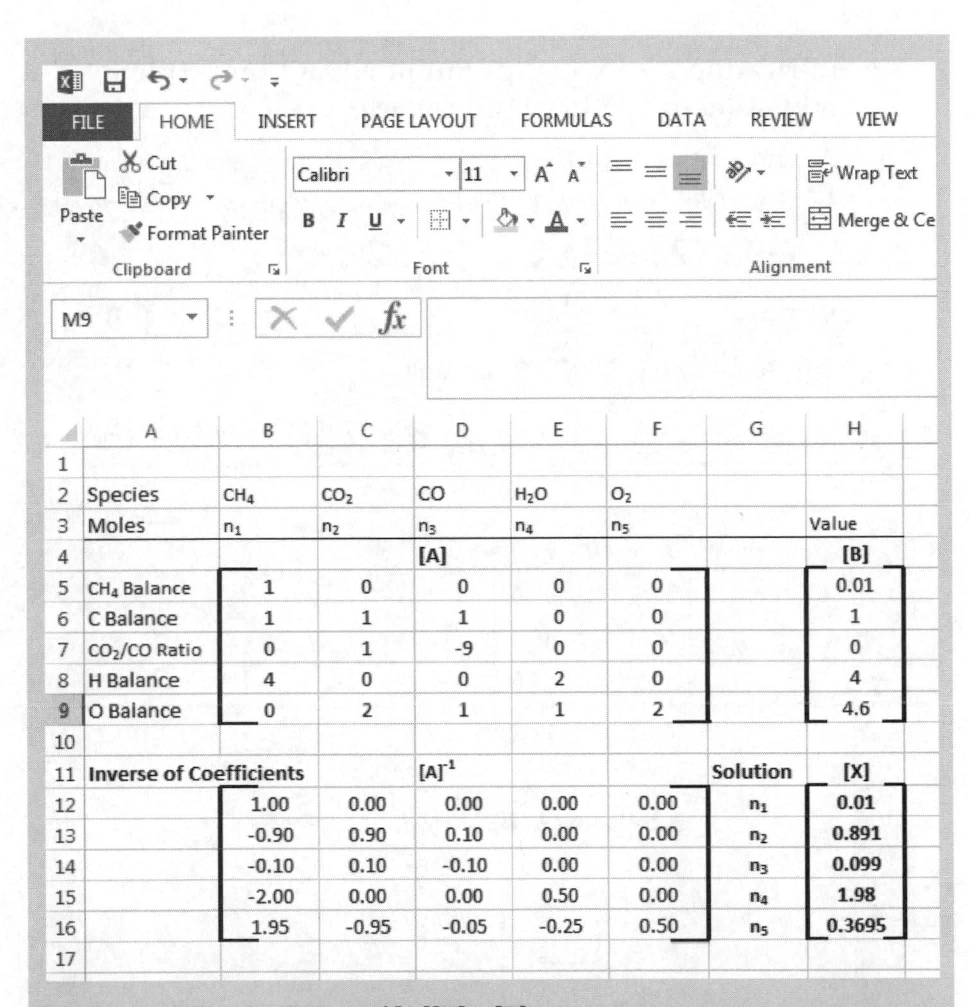

	A	B	C	D	E	F	G	H
1								
2	Species	CH_4	CO_2	CO	H_2O	O_2		
3	Moles	n_1	n_2	n_3	n_4	n_5		Value
4				[A]				[B]
5	CH_4 Balance	1	0	0	0	0		0.01
6	C Balance	1	1	1	0	0		1
7	CO_2/CO Ratio	0	1	-9	0	0		0
8	H Balance	4	0	0	2	0		4
9	O Balance	0	2	1	1	2		4.6
10								
11	**Inverse of Coefficients**			$[A]^{-1}$			**Solution**	**[X]**
12		1.00	0.00	0.00	0.00	0.00	n_1	**0.01**
13		-0.90	0.90	0.10	0.00	0.00	n_2	**0.891**
14		-0.10	0.10	-0.10	0.00	0.00	n_3	**0.099**
15		-2.00	0.00	0.00	0.50	0.00	n_4	**1.98**
16		1.95	-0.95	-0.05	-0.25	0.50	n_5	**0.3695**
17								

Figure 6.7 Excel solution of [A][X] = [B].

The calculation steps are as follows:

1. Enter the matrix coefficients by typing A:, and then either clicking on the matrix icon on the toolbar or pressing the <Ctrl+m> keys simultaneously. This brings up a dialog box; specify the number of rows and columns to be 5 each. This results in the creation of a 5 × 5 matrix with a placeholder for each matrix element. The values of the coefficients can now be entered by clicking on each placeholder.
2. Enter the matrix **B** using the same procedure, that is, by typing B: and pressing <Ctrl+m> or clicking on the matrix icon. The number

(Continues)

Modification 2 Excess Air Supplied and Incomplete Combustion of Methane (*Continued*)

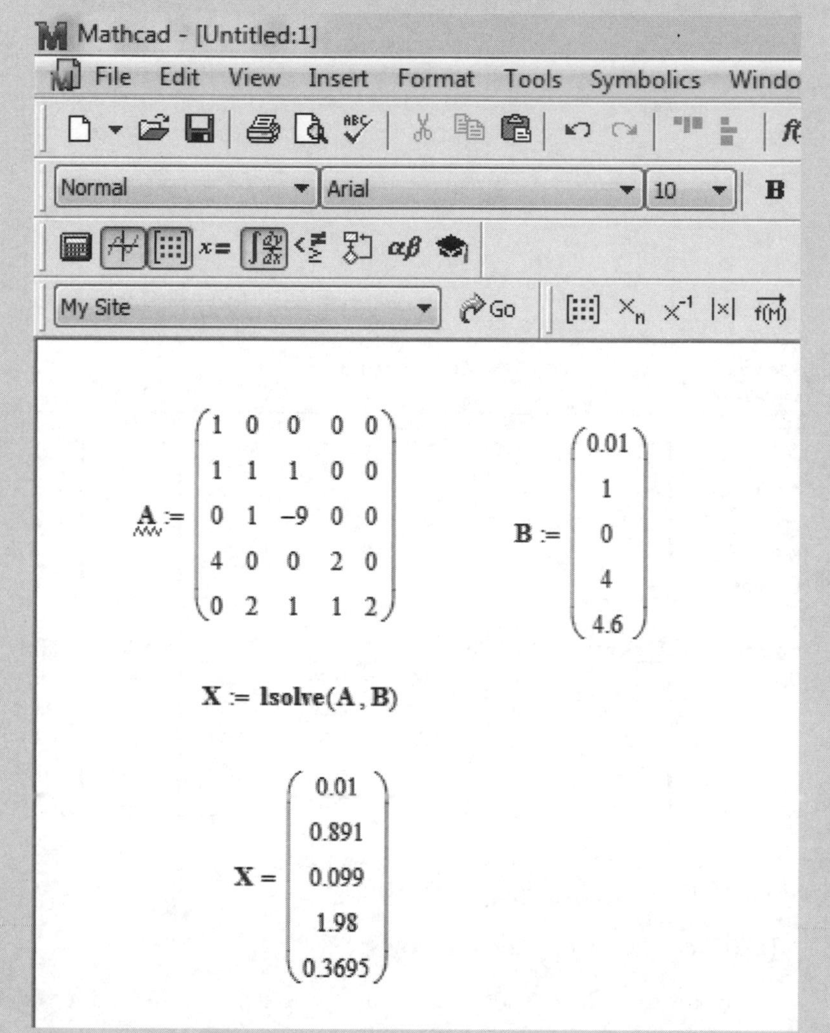

Figure 6.8 Mathcad solution of [A][X] = [B].

of columns is 1 and the number of rows 5. Appropriate values are entered for each element.

3. Solve the system by typing X:lsolve(**A**,**B**).
4. Type X= results to display the solution.

Alternatively, a solution can also be obtained by entering the expression **X**:**A**$^{-1}$***B**. The solution is displayed by typing X=. Confirming that an identical solution is obtained is an exercise left to the reader.

As can be seen from Figures 6.6, 6.7, and 6.8, all three techniques yield identical solutions. It should be noted that it is not necessary to resort to matrix computations to solve this particular problem. The first equation readily yields the value of n_1, as it involves only this variable. This value can then be substituted in other equations in appropriate sequence to solve other equations. Unfortunately, most practical systems will not yield such a convenient set of equations; however, the matrix computations illustrated here will be useful in arriving at the solution of the material balance problem.

Most practical reacting systems will involve, apart from the desired reaction, several side reactions of the primary reactant of interest. Further, the reactant mixture will inevitably contain several other constituents that undergo multiple reactions as well. The system may have multiple outlet streams in different physical states (gas, liquid, solid). It can be understood that the complexity of the system can increase significantly. Such systems cannot be solved easily through manual calculations using a basic scientific calculator (as for the system previously discussed), necessitating the use of software such as Excel or Mathcad that can rapidly solve systems containing several dozens (and possibly hundreds) of equations and variables.

6.4 Material Balances over Multiple Process Units

The incompleteness of the desired reaction and the occurrence of undesired side reaction introduced additional levels of complexity in the simple combustion process previously described. By now, readers should have developed a sense of the enormity of computational tasks for a chemical process plant, considering that a chemical process typically has a large number of interconnected process units handling complex streams. An example of such computations involving multiple units is presented in Example 6.4.1. The example deals with a (considerably) simplified ammonia (NH_3) synthesis process.

Example 6.4.1 Air, Methane, and Steam Requirement for Ammonia Synthesis

What are the quantities of the air, CH_4, and steam needed to produce 1000 tpd of NH_3?

Solution

NH_3 synthesis requires N_2 and hydrogen (H_2). As mentioned in previous chapters, H_2 needed for the process is typically obtained by steam

(*Continues*)

Example 6.4.1 Air, Methane, and Steam Requirement for Ammonia Synthesis (*Continued*)

reforming of CH_4, hence the need for CH_4 and steam. N_2 needed is obtained from air. We will assume in this example that N_2 is obtained by cryogenic distillation of air. The simplified block diagram of the process is shown in Figure 6.9.

Typically, the approach to solving the material balance computational problems involves first defining a basis for computation. The NH_3 production capacity of the reactor stated in the problem is 1000 tpd. However, computations are rarely based on such stated numbers. Typically, the problem is solved assuming a molar basis—in this case, the basis is 1 mole of NH_3 produced. The stepwise procedure beginning with this basis follows:

1. Compute the N_2 and H_2 requirements from the reaction stoichiometry ($0.5N_2 + 1.5H_2 \rightarrow NH_3$). Here the molar stoichiometric ratios are $H_2/NH_3 = 1.5$; $N_2/NH_3 = 0.5$. Therefore, the inlet to the reactor consists of 0.5 moles of N_2 and 1.5 moles of H_2.
2. Obtain the 0.5 moles of N_2 from the air separation unit, as shown in the figure. The total air fed to the unit must be $0.5/0.79 = 0.63$ moles. Assuming that the air is simply a mixture of O_2 and N_2, this results in 0.13 moles of O_2 being fed to the separation unit, and these are recovered as pure O_2 product.
3. Obtain the 1.5 moles of H_2 needed for the process from the steam reforming process. The steam reforming and water-gas shift reactions follow:

$$\text{Steam reforming: } CH_4 + H_2O \rightarrow CO + 3H_2$$

$$\text{Water-gas shift: } CO + H_2O \rightarrow CO_2 + H_2$$

$$\text{Overall: } CH_4 + 2H_2O \rightarrow CO_2 + 4H_2$$

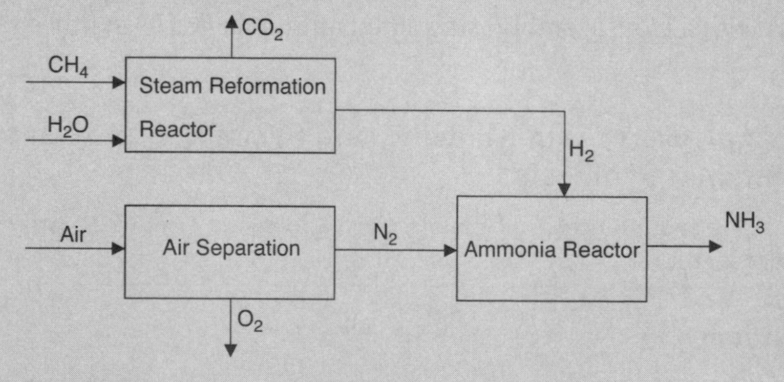

Figure 6.9 Simplified block flow diagram of ammonia synthesis process.

The overall stoichiometry of the process indicates that 4 moles of H_2 are obtained from 1 mole of CH_4 and 2 moles of H_2O (steam). Hence, the molar flow rates of CH_4 and H_2O needed are 1.5/4 and 1.5/2, that is, 0.375 and 0.75 moles, respectively. The process also generates 0.375 moles of CO_2.

4. Once the quantities of all the species have been determined on the chosen basis (1 mole of NH_3 produced), calculate the actual quantities for the specified capacity of 1000 tpd of NH_3—that is, 58,823 kilomoles of NH_3—by simply scaling by the mole ratio (58,823 kilomoles/mole), as shown here and in Figure 6.10:

Ammonia	1000 tpd	58,823 kilomoles
Methane	353 tpd	22,059 kilomoles
Water	794 tpd	44,118 kilomoles
Air	1075 tpd	37,058 kilomoles
Oxygen	249 tpd	7,782 kilomoles
Carbon dioxide	970 tpd	22,059 kilomoles

Based on these numbers, the total mass flow into the system (CH_4, air, H_2O) is 2,222 tpd, and the total outflow from the system (NH_3, O_2, CO_2) is 2,219 tpd. This discrepancy is due to rounding errors rather than any computational or analysis error.

Material flows for N_2 and H_2 are not shown in Figure 6.10. These calculations are left to the reader, who is also encouraged to confirm the material balance for each individual unit of the process. The dashed line in Figure 6.10 forms an envelope establishing a system boundary. The inlet and outlet mass flow rates crossing the boundary are balanced as shown. Similar envelopes can be drawn around any part of the process, and material balance for that part verified.

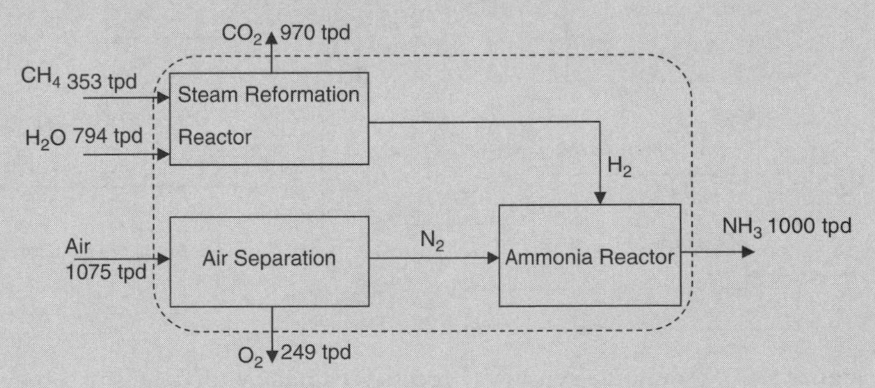

Figure 6.10 Material flows for ammonia synthesis process.

The steam reforming reaction is highly endothermic, and required elevated temperatures are achieved through the combustion of CH_4 itself. When stoichiometric quantity of air is used for the combustion of CH_4, CO_2 and H_2O (and other impurities) can be removed from the exhaust gases, yielding N_2 needed for NH_3 synthesis [5]. CO_2 and H_2O removal is typically carried out by absorption in a solvent, usually monoethanolamine [6]. The schematic flowsheet for this mode of operation is shown in Figure 6.11.

It should be noted that the steam reforming reaction also produces a stream containing CO_2, the other major component of that stream being H_2. CO_2 removal from this stream also requires separation operations such as absorption and membrane separation that were not shown in Figures 6.9 and 6.10.

The dashed lines in the steam reformation reactor in Figure 6.11 indicate that the two streams—the process stream composed of CH_4 and steam, and the combustion gas stream consisting of CH_4 and air—are physically isolated from each other. Typically, the process stream flows through tubes arranged inside a combustion chamber. The tubes are packed with the appropriate catalyst needed for the reaction.

The process shown in Figure 6.11 is considerably more complicated than the one shown in Figure 6.9. However, the actual process is even more complex, with several more separation units needed for removals/recoveries of other components and recycle streams for the reactor and other units. The principles of material balances for such complex processes and techniques

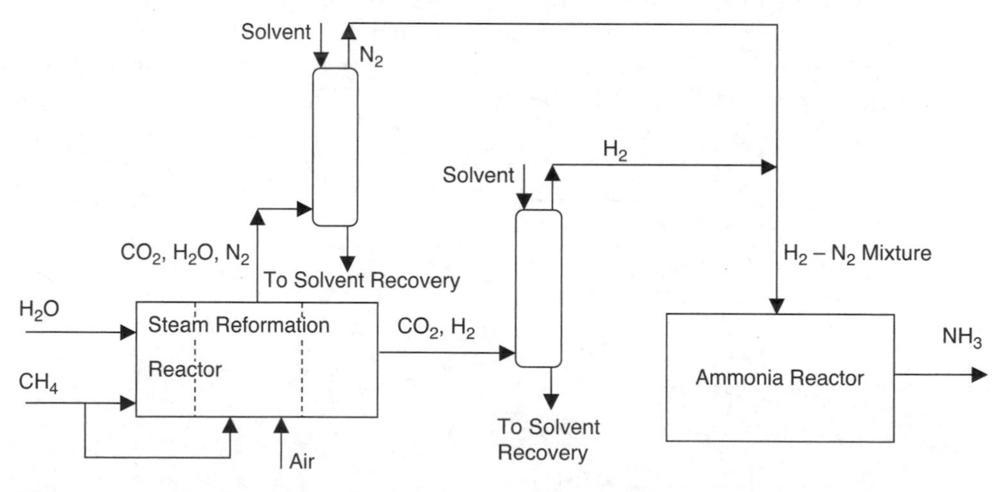

Figure 6.11 Ammonia synthesis process schematic with details of separation.

for performing the computations are taught in the Material and Energy Balance course taken in the sophomore year of the chemical engineering program, as explained in Chapter 3.

6.5 Summary

The principle of conservation of mass is employed by chemical engineers to perform the material balance computations on process units and plants. These computations involve the overall material balance and the balances for components present in the system. Molecular species (compounds) are conserved and hence serve as components in nonreacting systems, whereas for reacting systems, component balances involve atomic species. The material balance computations involve the simultaneous solution of linear algebraic equations, as illustrated by various examples in this chapter. Solution procedures using built-in capabilities of Excel and Mathcad were shown. Similar capabilities are offered by many other software systems, and choice of the particular tool to be used depends on its accessibility and availability to the user. The complexity of computational needs for performing material balances on multiple units and entire processes was also described. A chemical engineer will learn to use the tools for performing such comprehensive computations in the sophomore-level course on material and energy balances.

References

1. Hougen, O. A., K. M. Watson, and R. A. Ragatz, *Chemical Process Principles. Part I: Material and Energy Balances,* Second Edition, John Wiley and Sons, New York, 1958.
2. Reklaitis, G. V., *Introduction to Material and Energy Balances,* John Wiley and Sons, New York, 1983.
3. Himmelblau, D. M., and J. B. Riggs, *Basic Principles and Calculations in Chemical Engineering,* Eighth Edition, Prentice Hall, Upper Saddle River, New Jersey, 2012.
4. Felder, R. M., and R. W. Rousseau, *Elementary Principles of Chemical Processes,* Third Edition, John Wiley and Sons, New York, 2005.
5. Chenier, P. J., *Survey of Industrial Chemistry,* Third Edition, Springer, Berlin, Germany, 2002.
6. Thambimuthu, K., M. Soltanieh, J. C. Abanades, et al., "Capture of CO_2," *IPCC Special Report on Carbon Dioxide Capture and Storage,* edited by B. Metz,

O. Davidson, H. de Coninck, M. Loos, and L. Meyer, Chapter 3, Cambridge University Press, UK, 2005, https://www.ipcc.ch/pdf/special-reports/srccs/srccs_chapter3.pdf.

Problems

6.1 Raw French-cut potato strips for fries are dried in a conveyer-type dryer, which operates at 60°C. The H_2O content of the strips at the inlet to the dryer is 60% by mass. Dry air at 60°C is fed to the dryer at a volumetric flow rate of 1000 ft^3/min. Air exiting the dryer is saturated with water vapor with the saturation moisture content of 3.7 g H_2O per dry ft^3 of air. The feed rate of strips is 20 kg/min. What is the moisture content of strips exiting the dryer? What is the condensate flow rate if the air exiting the dryer is passed through a condenser for the removal of moisture?

6.2 A wastewater stream containing 4% (by mass) acetaldehyde (CH_3CHO, molar mass 44 g/mol) is contacted with N_2 in a stripping column to reduce the acetaldehyde concentration to 0.2% by mass. The wastewater stream flow rate at the inlet is 500 kg/hr. The N_2 exiting the column contains 10% acetaldehyde by volume (mole). What is the flow rate of acetaldehyde in the gas phase leaving the column? What is the flow rate of N_2 if the N_2 entering the column is acetaldehyde-free?

6.3 Elemental phosphorus is produced by heating phosphate ore with silica and CO. The reaction is

$$2Ca_3(PO_4)_2\,(s) + 6SiO_2(s) + 10C\,(s) \rightarrow 6CaSiO_3(s) + P_4(g) + 10CO(g)$$

How much carbon is needed to produce 1 tpd of phosphorus? What is the quantity of slag produced per year?

6.4 Rework example 6.3.1 (and its modifications) when the fuel burned is LPG (liquefied petroleum gas), treating it as an equimolar mixture of propane and butane. Following are the combustion reactions:

$$C_3H_8 + 5O_2 \rightarrow 3CO_2 + 4H_2O$$

$$C_4H_{10} + 6.5O_2 \rightarrow 4CO_2 + 5H_2O$$

$$C_3H_8 + 3.5O_2 \rightarrow 3CO + 4H_2O$$

$$C_4H_{10} + 4.5O_2 \rightarrow 4CO + 5H_2O$$

6.5 A gas-phase reaction between acetaldehyde and formaldehyde in the presence of a base results in the formation of pentaerythritol ($C(CH_2OH)_4$) according the following equation:

$$4HCHO + CH_3CHO + {}^-OH \rightarrow C(CH_2OH)_4 + HCOO^-$$

How much formaldehyde is needed per kilogram of pentaerythritol? What is the consumption rate of the base if the base used is $Ca(OH)_2$?

6.6 Electrolysis of salt (NaCl) solution in a diaphragm cell yields an ~30% (w/w) NaOH solution. How much H_2O needs to be evaporated per ton of product sold as 50% (w/w) solution? What is the quantity of 30% stream that needs to be processed?

6.7 The overall reaction for the electrolysis process in problem 6.6 can be represented as follows:

$$2NaCl(aq) + H_2O(l) \rightarrow 2NaOH(aq) + H_2(g) + Cl_2(g)$$

What are the quantities of chlorine (Cl_2) and H_2 produced per ton of 50% (w/w) NaOH solution?

6.8 The world burns 4000 million tons (oil equivalent) of coal per year. Assuming that oil equivalent coal is 78% carbon by mass, how many moles of CO_2 are emitted to the atmosphere each year from burning coal?

6.9 Natural uranium consists of 0.7% of fissile ^{235}U isotope, with the balance being the fertile ^{238}U isotope. Conventional light water reactors are typically refueled every 12 to 18 months by replacement of used fuel by fresh fuel enriched in the fissile isotope to 3% to 5%. How much natural uranium needs to be processed for refueling the reactor with 25 tons of uranium oxide (UO_2) fresh fuel enriched to 3.5% in ^{235}U, assuming that all the fissile isotope present in the natural uranium can be recovered in the enriched fuel? How does the answer change if the enrichment process also results in the formation of a reject (tailings) stream containing 0.1% ^{235}U? All concentrations are on a mass basis.

6.10 Ethanol is dehydrogenated to acetaldehyde by a vapor-phase catalytic reaction:

$$C_2H_5OH \rightarrow CH_3CHO + H_2$$

The conversion in the reaction is 35%, meaning that only 35% of the ethanol fed to the reactor undergoes dehydrogenation. H_2 is separated from the product mixture by cooling the product stream to condense both ethanol and acetaldehyde, which are then further separated by distillation for recovering the acetaldehyde product and recycling ethanol to the reactor. The condensation step is effective in condensing 99.9% of ethanol but only 98% of the acetaldehyde fed to it. Calculate the compositions of the H_2 product stream and the stream fed to the distillation column.

CHAPTER 7

Energy Balance Computations

The sum of actual and potential energies in the universe is unchangeable.

—William John Macquorn Rankine[1]

This quote by William Rankine is simply an alternative statement of the principle of the *conservation of energy* or the *first law of thermodynamics—that energy can be neither created nor destroyed*. This principle provides the foundation for the energy balance computations carried out by chemical engineers. As energy can be neither created nor destroyed, it follows that if an object or a system experiences a decrease in a certain form of energy, then this decrease must be compensated exactly by an increase in other forms of energy, including heat and work. Quantification of energy flows thus allows a chemical engineer to calculate temperatures of process streams and units, heat effects (evolution or absorption), and work obtained from the system or done on the system. Because energy and work always have a monetary value associated with them, it is as important for a chemical engineer to master the principles of energy balance to account for various forms of energy and interconversion among them as it is to account for the material in a chemical process. This chapter first presents the general principles of energy balance, then introduces some of the typical energy balance computations.

7.1 Quantitative Principles of Energy Balance

Because energy can take several different forms, it is necessary for us to understand the forms that are of primary interest to a chemical engineer. These forms are briefly described in the following section.

1. Nineteenth-century Scottish engineer and scientist who made fundamental contributions to the development of thermodynamics; inventor of the *Rankine cycle* for conversion of heat into mechanical energy/work. The Rankine temperature scale is named after him. Quotation source: Rankine, W. J. M., "On the general law of the transformation of energy," *Philosophical Magazine*, Vol. 5, No. 30, 1853, pp. 106–117.

7.1.1 Forms of Energy

Three primary forms of energy are encountered in a chemical process [1]:

1. *Kinetic energy* (*KE*)—Energy associated with motion. Kinetic energy of a body of mass m and velocity v is $\frac{1}{2}mv^2$. Clearly, a body moving at a higher velocity has higher kinetic energy than one of equal mass but lower velocity.
2. *Potential energy* (*PE*)—The energy associated with position. Potential energy of a body of mass m is mgh, where g is acceleration due to gravity and h, its distance from the earth's surface.
3. *Internal energy* (*U*)—Stored energy associated with atomic and molecular structure and characteristics.

Other forms of energy, such as electric and magnetic field energies, are typically not of interest in chemical processes [2] and are not included in the analysis presented in this book. The energy of a system, E, comprises these three primary forms. The first two forms of energy (kinetic and potential) constitute the *mechanical energy* of the system.

7.1.2 Generalized Energy Balance

Consider an arbitrary process represented by the block flow diagram shown in Figure 7.1.

This diagram is similar to the one shown in Figure 6.1 for the material balance, except for the following important differences: First, the streams represent energy flows rather than material flows. The solid lines indicate energy flows that accompany material flows into and out of the process unit. (All inlet/outlet energy flows that accompany the material flows are combined into one stream each.) Second, the system representing the process unit can exchange energy with the surroundings through either work or heat.[2] These streams are shown by dotted (work) and dashed (heat) lines, respectively.

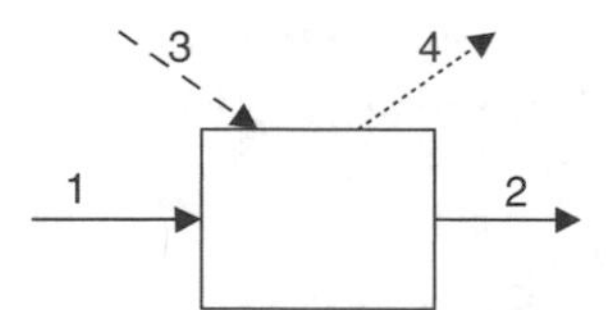

Figure 7.1 Energy balance on a process unit; Streams: 1—Inlet, 2—Outlet, 3—Heat, 4—Work.

2. The detailed discussion of concepts of work and heat is typically covered in the engineering thermodynamics courses and is not attempted here. Heat is simply understood to be the form of energy transferred from an object at a higher temperature to another at a lower temperature.

It is, of course, not necessary that there be an exchange of heat between the system and the surroundings. Similarly, the system may or may not do any work (or have work done on it). In the most general case, the application of the conservation of energy principle to this system leads to the following energy balance equation:

Rate of accumulation of energy in system =
Rate of energy flow in with material flows
– Rate of energy flow out with material flows
+ Rate of heat input to the system
– Rate of work done by the system

Note that the direction of heat and work streams will change depending on whether heat is added to or removed from the system, and work is done by the system or done on the system. Physically, the material being fed to the process must either accumulate in it or exit the system, as discussed in Chapter 6, "Material Balance Computations."

If $\dot{E}_1$ and $\dot{E}_2$ represent the rates of energy flowing in and out respectively along with the material flows, $\dot{Q}$ is the rate of heat input, and $\dot{W}$ is the rate of work done by the system, then we have the following:

$$\frac{dE}{dt} = \dot{E}_1 - \dot{E}_2 + \dot{Q} - \dot{W} \tag{7.1}$$

Here, E is the total energy of the system.

Equation 7.1 is the mathematical representation of the *overall energy balance* for the system or process. If the system is operating at steady state, then we have the following:

$$\dot{E}_1 - \dot{E}_2 + \dot{Q} - \dot{W} = 0 \tag{7.2}$$

There are several situations in chemical processes where no heat exchange is involved *and* internal energy changes are negligible. The energy balance then reduces to a *mechanical energy balance*, typically encountered in fluid flow situations and used for determining power requirements for transferring material from one point to another. On the other hand, the changes in mechanical energy are almost invariably insignificant as compared to the changes in internal energy in almost all other process units, and the energy balance simplifies to the following:

$$\Delta\dot{U} = \dot{U}_2 - \dot{U}_1 = \dot{Q} - \dot{W} \tag{7.3}$$

In this equation, $\Delta\dot{U}$ is the difference in internal energy between the outlet and inlet energy streams.

Equation 7.3 is useful for performing energy balance computations for steady-state processes. However, the equation is somewhat cumbersome and unwieldy to use and can be simplified further. The conceptual basis for this simplification is as follows:

The work input is typically considered to comprise three components—the flow work, the shaft work (such as that done by a pump or an agitator in a tank), and other work. The flow work terms associated with streams, when combined with the internal energy terms for the streams, yield the *enthalpies*[3] of the streams. Any other type of work excluding the flow and shaft work is lumped together as other work. If the shaft and other work can also be neglected, then the energy balance can be written as follows:

$$\dot{H}_2 - \dot{H}_1 = \dot{Q} \tag{7.4}$$

Equation 7.4 is simply stating that the difference between the heat contents of outlet and inlet streams is equal to the heat supplied to the system.

Several processes operate under *adiabatic* conditions, that is, there is no heat exchange between the system and the surroundings ($\dot{Q} = 0$). The energy balance for these processes is written as follows:

$$\dot{H}_2 - \dot{H}_1 = 0 \tag{7.5}$$

Equations 7.4 and 7.5 provide the framework for the energy balance computations for chemical processes. However, these equations are in terms of enthalpy, a thermodynamic quantity, which cannot be measured and whose absolute value for a substance is unknown. Practical application of energy balances requires expressing the equations in terms of measurable quantities, such as temperature. The next section presents some basic principles related to enthalpy and illustrates how the previous equations are converted into useful forms. The rigorous mathematical development of the simplifying steps is left to the thermodynamics and material and energy balance courses.

7.1.3 Enthalpy and Heat Capacity

Enthalpy is a measure of the energy (or heat) content of a substance [3]. It is a thermodynamic quantity whose *absolute* value cannot be determined; however, enthalpy of a substance with respect to its value at some reference conditions can be calculated [4, 5]. The reference state, also called the *standard state,* is

3. Enthalpy (H) is a thermodynamic quantity that is a measure of the heat content of a stream or object: $H = U + PV$; P is the pressure, V is the volume.

specified in terms of pressure and temperature of the system, usually 1 bar and 25°C (298.15 K) [6]. The standard specific enthalpies (enthalpy per mole) of formation of various substances (from its constituent elements) at the reference state are available from various sources, including books on thermodynamics [3], handbooks [7], and Web databases such the one maintained by the National Institute of Standards and Technology (www.nist.gov).[4] Thus, the specific enthalpy of any substance at any other condition can be calculated from its functional dependence on system variables and the reference state enthalpy.

Enthalpy is a function of temperature and pressure of the system, and its dependence on temperature at constant pressure is described as follows:

$$\left(\frac{\partial h}{\partial T}\right)_P = C_P \tag{7.6}$$

The left side of this equation represents the partial derivative of enthalpy with respect to temperature at constant pressure; h is the specific enthalpy of the substance—that is, enthalpy per unit mole—and C_P is the specific heat capacity of the substance at constant pressure. The SI units of h and C_P are joule per mole (J/mol) and joule per mole per kelvin (J/mol K), respectively. The specific enthalpy h does not depend on the quantity of substance present, making it an *intensive* property. The total enthalpy H, on the other hand, is an *extensive* property, which depends on the quantity of material present in the system.[5] H is obtained simply by multiplying the specific enthalpy by the number of moles present and has the unit of J (joule).

If the information about the specific heat capacity C_P is available, then integration of equation 7.6 enables us to calculate the change in enthalpy (Δh) when the temperature of the substance changes from T_1 to T_2 at constant pressure:

$$h_2 - h_1 = \Delta h = C_P\left(T_2 - T_1\right) \tag{7.7}$$

where h_1 and h_2 are the specific enthalpies of the substance at the temperatures T_1 and T_2, respectively.

Note that equation 7.7 is valid only when C_P, the specific heat capacity at constant pressure, does not depend on the temperature and hence is constant

4. By convention, the enthalpies of formation of elements in their natural states of occurrence are taken to be zero.

5. The intensive and extensive properties are discussed in detail in the thermodynamics courses.

over the temperature range under consideration. Typically, however, C_P is a function of temperature, the dependence often being expressed as polynomial in T, with one such function shown by equation 7.8 [5].

$$C_P = A + BT + CT^2 + DT^{-2} + ET^3 \quad (7.8)$$

Coefficients A through E are constants characteristic of the substance and are available from the same sources previously stated. The enthalpy change per unit mole of the substance is then calculated by integrating equation 7.6:

$$\Delta h = \int_{T_1}^{T_2} C_P \, dT \quad (7.9)$$

Equation 7.7 or 7.9 is used for calculating the change in the specific enthalpy of a substance when its temperature changes from T_1 to T_2 under constant pressure conditions. When the process is not conducted under constant pressure (isobaric) conditions, enthalpy dependence on pressure also needs to be taken into account while performing the energy balance computations. The pressure dependence of enthalpy is complex and requires an understanding of the volumetric behavior of the substances—that is, an understanding of the relationship between pressure, volume, and temperature for the substance. This is generally covered in the thermodynamics courses and is not considered in this text.

The assumption implicit in the development of equations 7.7 and 7.9 is that the substance *does not undergo a phase change*; that is, it does not change its state from solid to liquid or liquid to gas, and vice versa. Thus, the substance undergoes only a *sensible heat change* that is reflected in the temperature of the substance. However, if the substance does experience a phase change at a temperature intermediate between T_1 and T_2, then the enthalpy change should include a *latent heat* component. For example, if the boiling point of the substance T_b is greater than T_1 but less than T_2, then the substance is a liquid at the beginning of the process at T_1, but at T_2, at the end of the process, it is a vapor. The enthalpy change for this situation is described by equation 7.10.

$$\Delta h = \int_{T_1}^{T_b} C_{P_L} \, dT + \Delta H_v + \int_{T_b}^{T_2} C_{P_V} \, dT \quad (7.10)$$

In this equation, C_{P_L} and C_{P_V} are the specific heat capacities of the liquid and vapor form of the substance, respectively, and ΔH_v is the latent heat of vaporization at the temperature T_b. If the phase change involves melting/fusion or sublimation/condensation, then the corresponding latent heat value must be used.

If T_1 is chosen as 298.15 K—that is, the standard state temperature—then the specific enthalpy of a substance can be calculated at any temperature using equation 7.11.

$$h(T) = h_{298.15} + \Delta h = \Delta H_F^0 + \Delta h \tag{7.11}$$

Here, Δh is calculated using equation 7.9 or 7.10, with the lower and upper temperature limits of integration being 298.15 and T K, respectively. The specific enthalpy at 298.15 K, $h_{298.15}$, is equal to the standard enthalpy of formation, ΔH_F^0, as previously discussed. Equation 7.11 allows us to compute the specific enthalpy of any substance at any temperature, provided the information on the standard enthalpy of formation and the dependence of the specific heat capacity on temperature are known.

7.1.4 Enthalpy Changes in Processes

The previous discussion should make it clear that it is possible to obtain the values of specific enthalpy of any substance at any temperature. It follows that if a process is carried out at a certain temperature—that is, both the feed and product streams are at that specified temperature—then a certain enthalpy change is associated with that process. The following generic reaction is an example:

$$\mathrm{A} + (b/a)\,\mathrm{B} \rightarrow (c/a)\,\mathrm{C} + (d/a)\,\mathrm{D}$$

The enthalpy of reaction (or heat of reaction) is simply the difference between the enthalpies of products and enthalpies of reactants. The following shows this mathematically:

$$\Delta H_{rxn} = \sum H_{products} - \sum H_{reactants} = \sum_i \left(v_i h_i\right)_{products} - \sum_j \left(v_j h_j\right)_{reactants} \tag{7.12}$$

Here, ν represents the stoichiometric coefficients of the species involved in the reaction. It should be noted that the equation for the reaction is written such that the stoichiometric coefficients of all the other species are normalized with respect to the stoichiometric coefficient of A; that is, the equation involves 1 mole of A and proportional moles of other species. Thus the enthalpy or heat of reaction, ΔH_{rxn} is based on 1 mole of reactant A. Of course, the equation can be normalized on the basis of the stoichiometric coefficient of any other species involved in the reaction, with the enthalpy of the reaction changing proportionately.

If the process is conducted at standard conditions, then the enthalpy change is termed as the *standard enthalpy change*. For the reaction shown previously, the standard enthalpy of reaction follows:

$$\Delta H^0_{rxn} = \left(\frac{c}{a} \Delta H^0_{F,C} + \frac{d}{a} \Delta H^0_{F,D} \right) - \left(\frac{b}{a} \Delta H^0_{F,B} + \Delta H^0_{F,A} \right) \tag{7.13}$$

If the standard enthalpies of the reactants are higher than those of the products, then the enthalpy of the reaction will be negative. The process involves starting with a material having higher chemical energy and ending up with a material with a lower energy. The difference between the two energies (or enthalpies) appears as the heat evolves during the transformation, making the process *exothermic*. Conversely, if standard enthalpies of the products are higher than those of the reactants, then the process involves starting with a material of lower energy and ending up with a material having higher energy. Such processes are termed *endothermic*. Figure 7.2 shows a conceptual schematic of the enthalpy changes in these two types of processes.

It is obvious that for an exothermic process, a mechanism for removing heat is necessary if it is desired to maintain a constant temperature. However, if the process is conducted adiabatically—that is, the system does not exchange heat with the surroundings—then the products will be at a higher temperature than the reactants. Conversely, if the process is endothermic, it will require heat input to maintain a constant temperature, and the adiabatic endothermic process will experience a decrease in temperature. Figure 7.3 shows the changes in temperature for an adiabatic system for both endothermic and exothermic processes.

When the transformation involves a chemical reaction, the enthalpy effect is termed the *enthalpy of reaction* or the *heat of reaction*. The enthalpy (or heat) of reaction is termed *enthalpy (or heat) of combustion* when the reaction is of

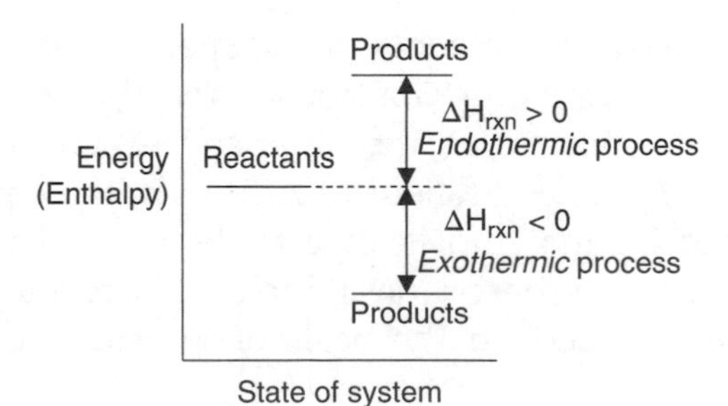

Figure 7.2 Conceptual schematic of enthalpy changes in endothermic and exothermic processes.

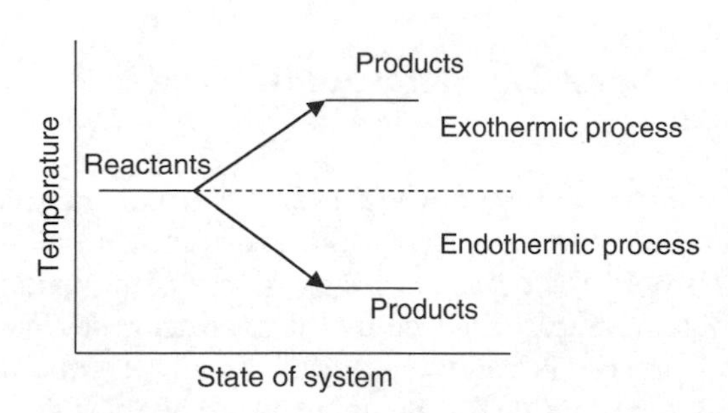

Figure 7.3 Heat effects in transformations: temperature of adiabatic systems.

combustion of a substance. Transformations that are physical in nature—that is, transformations that do not involve chemical reactions—are also frequently (usually) accompanied by an enthalpy change. For example, heat effects accompany dissolution of a solute in a solution, and the change in enthalpy is termed *enthalpy of solution* or *heat of solution*. Similarly, *enthalpy of mixing* refers to the enthalpy change when the process involves mixing of different streams. These transformations can be endothermic or exothermic as well. In all these cases, the discussion presented above for reactive systems can be extended, *mutatis mutandis*, to other processes and transformations.

7.2 Basic Energy Balance Problems

A chemical engineer has to perform a wide variety of energy balance computations for a large number of transformations and processes. These computations require application of the principles discussed in section 7.1. The myriad energy balance computations performed by chemical engineers can very broadly be classified into two types: those involving determination of heat effects in particular transformations of interest and those involving determinations of system temperatures as a result of the particular transformation undergone by the system. The following examples provide a brief glimpse into the nature of energy balance computations for different chemical processes.

Example 7.2.1 Adiabatic Flame Temperature of Exhaust Gases

Consider again the material balance discussed in example 6.3.1, the combustion of natural gas. Burning the natural gas releases heat—a manifestation of the principle of the conservation of energy, where the chemical energy stored

(*Continues*)

Example 7.2.1 Adiabatic Flame Temperature of Exhaust Gases (*Continued*)

in methane is converted into chemical energy of products and thermal energy. If this heat is not removed from the system, then it would cause the temperature of the product stream to rise. *Adiabatic flame temperature* in a combustion process is simply the theoretical temperature reached by the product gases in a combustion process, when the process is conducted adiabatically—without any energy exchange with the surroundings. Calculate the adiabatic flame temperature for complete combustion of natural gas when stoichiometric quantity of air is fed to the burner. All the inlet streams are at 25°C.

Solution

The first step in performing energy balance computations is obtaining the material balance for the process. These material balance calculations were conducted in Chapter 6 and are reproduced in Figure 7.4.

Note that Figure 7.4 shows the material balance for a continuous process where 1 mol/s of the natural gas is fed to the burner continuously. Since the process is adiabatic, at steady state the enthalpy flowing into the burner must equal the enthalpy leaving the burner:

$$\dot{H}_{out} - \dot{H}_{in} = 0 \quad \text{(E7.1)}$$

The enthalpy flowing in is equal to the sum of enthalpies carried into the burner by the three components:

$$\dot{H}_{in} = \sum_{i=1}^{3} \dot{n}_i h_i = \dot{n}_{CH_4} h_{CH_4} + \dot{n}_{O_2} h_{O_2} + \dot{n}_{N_2} h_{N_2} \quad \text{(E7.2)}$$

Here, $\dot{n}$ represents the molar flow rate of component i, and h_i its specific enthalpy (enthalpy per mole).

Similarly, the enthalpy flowing out of the burner is the sum of enthalpies of the components exiting the burner:

$$\dot{H}_{out} = \sum_{j=1}^{3} \dot{n}_j h_j = \dot{n}_{CO_2} h_{CO_2} + \dot{n}_{H_2O} h_{H_2O} + \dot{n}_{N_2} h_{N_2} \quad \text{(E7.3)}$$

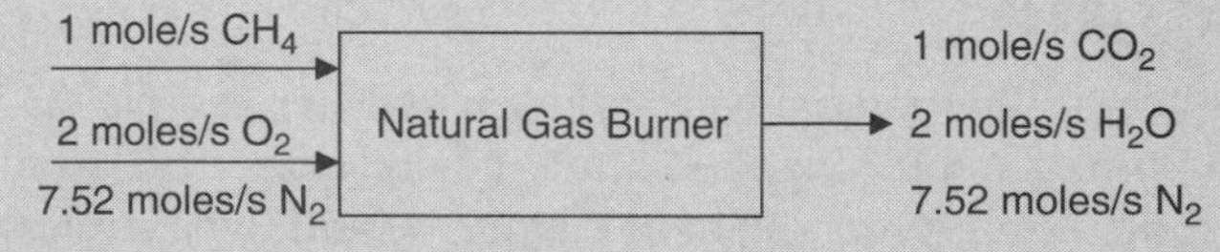

Figure 7.4 Material balance on a natural gas burner.

Enthalpies in equations E7.2 and E7.3 are evaluated at the corresponding stream temperatures; that is, at 25°C (298.15 K) for the inlet streams and an unknown temperature T (K) for the product stream. As mentioned earlier, enthalpies at 298.15 K are the enthalpies of formation at the standard state and are available from various sources. The enthalpies h_j can be expressed in terms of this temperature T, making use of equation E7.4:

$$h_{j,T_2} = h_{j,298.15} + \int_{298.15}^{T_2} C_{P_j}\, dT \qquad \text{(E7.4)}$$

Table E7.1 shows the standard enthalpies of formation ($\Delta H_F^0 = h_{298.15}$), and Table E7.2 shows the functional dependence of the heat capacity for each component [3, 7].

Using this data, the rate of enthalpy fed to the system ($\dot{H}_{in}$) is found to be equal to –74.87 kJ/s. The rate at which enthalpy is leaving the system is calculated as follows:

Table E7.1 Standard Enthalpies of Formation

Component	Enthalpy, kJ/mol
CH_4	–74.87
O_2	0
N_2	0
CO_2	–393.61
$H_2O(g)$[6]	–241.88

Table E7.2 Functional Dependence of Heat Capacity

Specific Heat Capacity $C_p = A + BT + CT^2 + DT^{-2}$ **J/mol K**

Component	A	B ($\times 10^3$)	C ($\times 10^6$)	D ($\times 10^{-5}$)
CH_4	14.15	75.5	–18.0	—
O_2	30.25	4.21	—	–1.89
N_2	27.27	4.93	—	0.33
CO_2	45.37	8.69	—	–9.62
H_2O	28.85	12.06	—	1.00

(Continues)

6. Gas phase.

Example 7.2.1 Adiabatic Flame Temperature of Exhaust Gases (*Continued*)

First, the specific enthalpy of each species at temperature T is calculated from the C_P data:

$$h_{j,T} = h_{j,298.15} + A_j(T - 298.15) + \frac{B_j}{2}(T^2 - 298.15^2) + \frac{C_j}{3}(T^3 - 298.15^3) - D_j(T^{-1} - 298.15^{-1}) \tag{E7.5}$$

Substituting equation E7.5 in equation E7.3 and rearranging terms gives us the following:

$$\begin{aligned}\dot{H}_{out} = \sum_{j=1}^{3} \dot{n}_j h_j = &\left(\dot{n}_{CO_2} h_{CO_2,298.15} + \dot{n}_{H_2O} h_{H_2O,298.15} + \dot{n}_{N_2} h_{N_2,298.15}\right) \\ &+\left(\frac{T^3 - 298.15^3}{3}\right)\left(\dot{n}_{CO_2} C_{CO_2} + \dot{n}_{H_2O} C_{H_2O} + \dot{n}_{N_2} C_{N_2}\right) \\ &+\left(\frac{T^2 - 298.15^2}{2}\right)\left(\dot{n}_{CO_2} B_{CO_2} + \dot{n}_{H_2O} B_{H_2O} + \dot{n}_{N_2} B_{N_2}\right) \\ &+(T - 298.15)\left(\dot{n}_{CO_2} A_{CO_2} + \dot{n}_{H_2O} A_{H_2O} + \dot{n}_{N_2} A_{N_2}\right) \\ &-(T^{-1} - 298.15^{-1})\left(\dot{n}_{CO_2} D_{CO_2} + \dot{n}_{H_2O} D_{H_2O} + \dot{n}_{N_2} D_{N_2}\right)\end{aligned} \tag{E7.6}$$

Equation E7.6 appears complicated, but that is primarily because all the individual terms are written in expanded form rather than because of any intrinsic complexity.

The molar flow rates of components in the outlet stream are as shown in Figure 7.4. Equating $\dot{H}_{out}$ to –74.87 kJ/s, and substituting the values of various parameters yields the following equation in unknown outlet temperature T:

$$0.03494T^2 + 308.14T - \frac{5.1384 \times 10^5}{T} - 899202 = 0 \tag{E7.7}$$

Equation E7.7 is a polynomial equation that can be solved using a number of different techniques. The solution using the Goal Seek tool in Microsoft Excel (described in Chapter 5, "Computations in Fluid Flow") is shown in Figure 7.5.

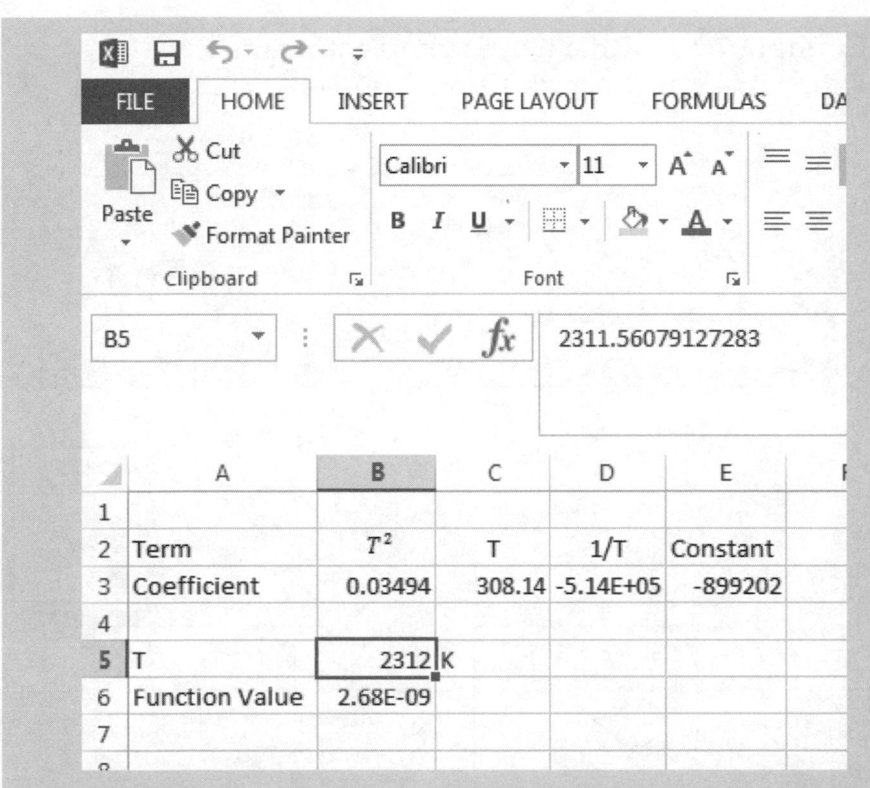

Figure 7.5 Excel Goal Seek solution for adiabatic flame temperature problem.

As seen from the figure, the coefficients of the terms T^2, T, and $1/T$ are entered in cells B3, C3, and D3, with the constant value in the cell E3. A temperature guess is provided in cell B5, and the function described by equation E7.7 is evaluated in cell B6. The Goal Seek function manipulates the value in B5 until the function value in B6 is approximately zero. The adiabatic flame temperature is 2312 K (~2039°C).

The graphical solution for the problem using Mathcad is shown in Figure 7.6.

The steps for obtaining the solution are as follows:

1. First, define a range for variable T by typing T:298.15;2500.
2. Define the function f(T) by typing

   f(T):0.03494*T^2 +308.14*T-5.1384*10^5/T-899202

(Continues)

Example 7.2.1 Adiabatic Flame Temperature of Exhaust Gases (*Continued*)

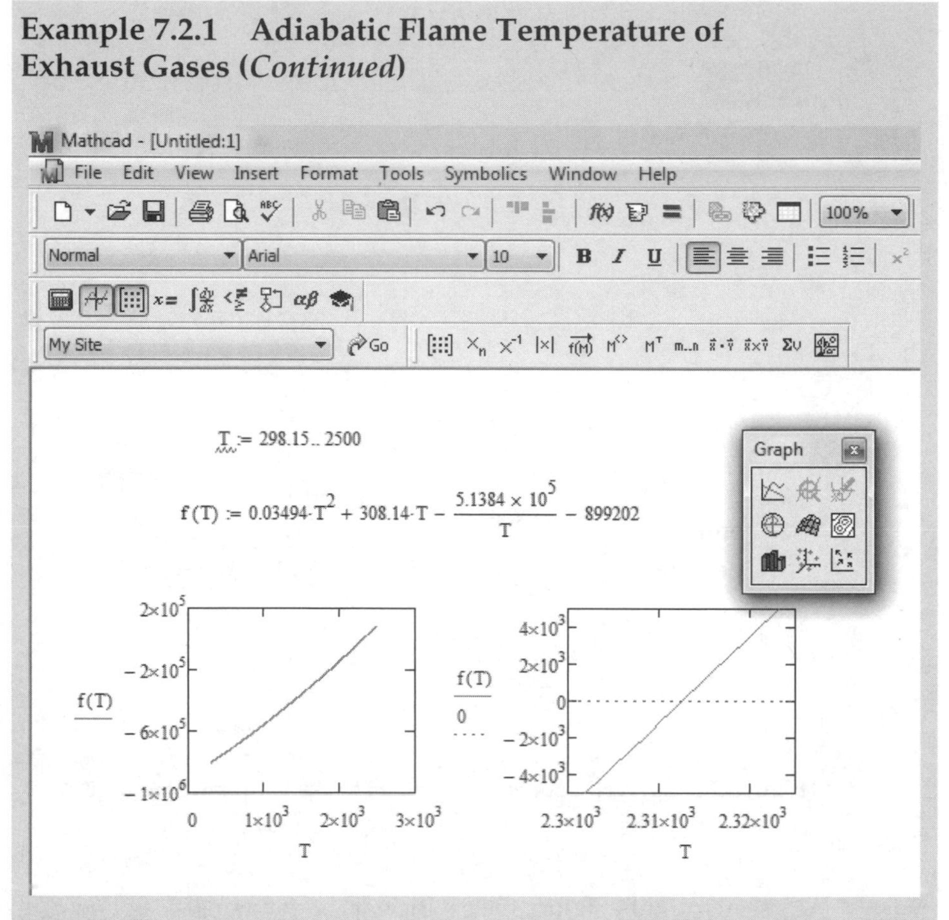

Figure 7.6 Mathcad graphical solution of the adiabatic flame temperature problem.

3. Create the graph of f(T) as a function of T by choosing Insert from the command menu, and then Graph and X-Y Plot from the drop-down menus. (Alternatively, a graph can also be created by clicking on the appropriate buttons.) This creates a blank graph with empty placeholders for the x and y variables. Enter T and f(T) by clicking on these placeholders, generating the graph shown at the bottom on the left. As can be seen from the graph, the function value goes from ~-8×10^5 to ~1.6×10^5, indicating that function value is zero in the vicinity of T = 2000. To get an accurate estimate, replot the function, as shown on the right, with the x-axis values between 2300 and 2325. In addition, add a straight horizontal line at y = 0 by clicking at the end of y-axis label f(T) and typing a space followed by 0.

4. The graph on the right clearly shows that the function is zero (from the intersection of the function curve with the horizontal dashed line) at a T value between by 2310 and 2315, enabling a guess for the solution at 2312. Improve the accuracy of the solution by progressively rescaling the x-axis to narrow the limits. However, it should be noted that a solution guess of 2310 is already within 0.1% of the exact solution and is adequate for most practical purposes.

Example 7.2.2 Heat Recovery from Combustion Gases

The combustion gases from the burner in the previous example are cooled to 25°C, in the process heating up water to 150°F (as would happen in a water heater). What is the flow rate of hot water if the cold water entering the heater is at 68°F?

Solution

The schematic of the process is shown in Figure 7.7.

As seen from the figure, the water is passed through the inside of the pipe, which is shaped into a coil to provide the necessary area for heat exchange. The energy balance on the process yields the following equation:

$$\dot{H}_{out,CG} + \dot{H}_{out,water} - \left(\dot{H}_{in.CG} + \dot{H}_{in.water}\right) = 0 \quad \text{(E7.8)}$$

CG stands for combustion gases. Rearranging the terms in Equation E7.8 gives us the following:

$$\dot{H}_{out,water} - \dot{H}_{in.water} = \dot{H}_{in.CG} - \dot{H}_{out,CG} \quad \text{(E7.9)}$$

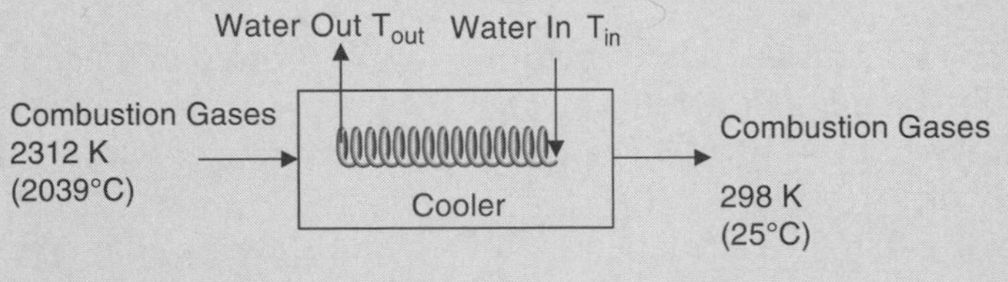

Figure 7.7 Energy balance on a water heater.

(*Continues*)

Example 7.2.2 Heat Recovery from Combustion Gases (*Continued*)

Equation E7.9 is the mathematical expression stating that the rate of enthalpy change for combustion gases is exactly equal to the rate of enthalpy change for water; that is, all the energy lost by the combustion gases is transferred to water. An assumption implicit in this statement is that the heater is perfectly insulated and there is no heat loss to the surroundings. The rate of enthalpy flowing in with the combustion gases, $\dot{H}_{in.CG}$, is −74.87 kJ/s, using the information from Example 7.2.1. The other terms in the equation E7.9 are calculated as follows:

$$\dot{H}_{out,CG} = \dot{n}_{CO_2} h_{CO_2,298.15} + \dot{n}_{H_2O} h_{H_2O,298.15} + \dot{n}_{N_2} h_{N_2,298.15} \quad \text{(E7.10)}$$

$$\dot{H}_{in,water} = \dot{n}_{water}\left(h_{water,298.15} + \int_{298.15}^{T_{In}} C_{P,water}\, dT\right) \quad \text{(E7.11)}$$

$$\dot{H}_{out,water} = \dot{n}_{water}\left(h_{water,298.15} + \int_{298.15}^{T_{Out}} C_{P,water}\, dT\right) \quad \text{(E7.12)}$$

From equation E7.11 and E7.12, the right side of equation E7.9 is simplified:

$$\dot{H}_{out,water} - \dot{H}_{in,water} = \dot{n}_{water}\left(\int_{T_{In}}^{T_{Out}} C_{P,water}\, dT\right) \quad \text{(E7.13)}$$

Using the enthalpies of formation data from Example 7.2.1, the left side of equation E7.9 is found to be 802.5 kJ/s. Assuming a constant specific heat capacity of 4.185 J/g K for water, the energy balance simplifies to the following:

$$\dot{n}_{water} \cdot 4.185 \cdot (338.7 - 298.15) = 802500 \quad \text{(E7.14)}$$

The inlet and outlet temperatures are expressed in the units of K rather than °F. The mass flow rate of water, $\dot{n}_{water}$, is readily calculated from equation E7.14 to be 4210 g/s. The volumetric flow rate of water is 4.21 L/s at a density of 1 g/cm^3 (1 kg/L).

Note that the calculations in this example are based on the assumption that all the components in outlet combustion gases remain in the gaseous state at 25°C. While this assumption is valid for nitrogen and carbon dioxide, it does not hold true of water, which will condense into liquid when cooled below 100°C at atmospheric pressure. If it is assumed that all of the water vapor present in the combustion gases condenses at 25°C,[7] then the standard enthalpy of formation of liquid water, −285.8 kJ/mol, should be used in the calculations instead of −241.88 kJ/mol for water vapor. This yields a water flow rate of 4671 g/s, or 4.67 L/s, as the answer. The reader is left to work out the computations to confirm the result.

Example 7.2.3 Calculation of Heat Transfer Area[8]

How much heat transfer area needs to be provided in order to achieve the desired outlet temperatures in the above problem?

Theoretical background

Heat exchange equipment is an integral and ubiquitous component in chemical process industries. Many process streams need to be heated to desired temperatures for conducting reactions and effecting separations. Several other process streams need to be cooled from high temperatures for proper handling of product streams and emissions to the environment. Most of this heat exchange requirement is met through recuperators—heat exchangers in which a flowing hot stream transfers heat to a cold stream across a barrier that keeps the two streams physically separated [8]. The driving force for this heat transfer is the temperature difference between the two streams. The thermal duty of the heat exchanger, $\dot{Q}$, is related to the driving force:

$$\dot{Q} = U \cdot A \cdot \text{Temperature Driving Force} \tag{7.14}$$

Here, A is the heat transfer area, and U is the overall heat transfer coefficient.

(Continues)

7. In reality, not all of the water vapor will condense. This situation is dealt with in Chapter 8, "Computations in Chemical Engineering Thermodynamics."

8. Calculations related to heat transfer are covered in the engineering thermodynamics and transport phenomena courses and typically are not covered in the sophomore-level material and energy balance course.

Example 7.2.3 Calculation of Heat Transfer Area (*Continued*)

The overall heat transfer coefficient U is a measure of rate at which heat is transferred across a surface. A higher value of U implies faster heat transfer, and consequently the area needed is lower. U depends on the properties of the fluid and the operating conditions—mainly the fluid velocity. The temperature driving force depends on the difference in the temperatures of the two fluids between which heat is transferred. This driving force varies with the location within the heat exchanger as the temperatures of both fluids are changing. The overall driving force is then obtained by taking a logarithmic mean of the temperature differences at the inlet and the outlet of the heat exchanger.[9] For the situation shown in Figure 7.7, the log-mean temperature difference (LMTD or ΔT_{LM}) is given by equation 7.15:

$$\Delta T_{LM} = \frac{\left(T_{CG,In} - T_{Water,Out}\right) - \left(T_{CG,Out} - T_{Water,In}\right)}{\ln\left(\dfrac{T_{CG,In} - T_{Water,Out}}{T_{CG,Out} - T_{Water,In}}\right)} \tag{7.15}$$

Solution

Rearranging equation 7.14 gives us the following equation:

$$A = \frac{\dot{Q}}{U \cdot \Delta\ T_{LM}} \tag{7.16}$$

The thermal duty of the heat exchanger, $\dot{Q}$, was found earlier in example 7.2.2 to be 802500 W (J/s). The temperature driving force, ΔT_{LM}, follows:

$$\Delta T_{LM} = \frac{(2312 - 338.7) - (298 - 293)}{\ln\left(\dfrac{2312 - 338.7}{298 - 293}\right)} = 329K$$

The overall heat transfer coefficient, as previously mentioned, depends on the properties of the fluids and the operating conditions. Various

9. The development of the concept of the log-mean temperature difference is discussed in engineering thermodynamics and transport phenomena courses.

empirical and semi-empirical correlations are available in various literature for the estimation of the heat transfer coefficients under different combinations of cold and hot fluids, types of operation, and so on. Approximate values of typical heat transfer coefficients when the detailed information about the fluids and operating conditions is not available are also available in literature. Using a representative value of 300 $W/m^2 K$ for U, the area needed is found to be 8.13 m^2. If the water is flowing through a 1 in. diameter pipe, then the length of the pipe needed to obtain this area is $8.13/(\pi \cdot 0.0254)$,[10] almost 101 m.

Example 7.2.4 Heat Effects in a Chemical Reaction

Calcium oxide is mixed with water in a process called *slaking* to obtain calcium hydroxide. In a batch tank at 25°C, 3000 kg CaO are mixed with 3000 kg water. What is the heat evolved in the reaction? The standard enthalpies of formation are CaO(s): –635.1 kJ/mol; water: –285.8 kJ/mol; and $Ca(OH)_2$ (aq): –1002.8 kJ/mol. The molar masses (g/mol) of the three substances are 56.1, 18.0, and 74.1, respectively.

Solution

The process is represented by the following equation:

$$CaO(s) + H_2O \rightarrow Ca(OH)_2\ (aq)$$

The enthalpy change in the reaction per mole of CaO follows:

$$\Delta H_{rxn} = \Delta H^0_{F,Ca(OH)_2} - \left(\Delta H^0_{F,CaO} + \Delta H^0_{F,H_2O}\right) \quad \text{(E7.15)}$$

$$\Delta H_{rxn} = -1002.8 - (-635.1 - 285.8) = -81.9 \text{ kJ/mol CaO}$$

The negative sign associated with the enthalpy (heat) of the reaction indicates that the product has lower energy compared to the reactants. The difference between the two enthalpies appears as heat that is rejected to the surroundings, making this an exothermic process. For every mole of CaO converted to $Ca(OH)_2$, 81.9 kJ heat is evolved. The total heat evolved is obtained by multiplying it by the total number of moles of CaO fed to the process:

$$\text{Heat evolved} = 81.9 \text{ kJ/mol} \cdot 3000 \text{ kg} \cdot 1000 \text{ g/kg} / 56.1 \text{ g/mol} = 4380 \text{ MJ}$$

This is a substantially large quantity of heat, which, if not removed from the system, will result in an increase in the temperature of the solution.

(*Continues*)

10. Heat transfer takes place across the surface area of the pipe, which is equal to πdL for a pipe having a diameter d and length L.

Example 7.2.4 Heat Effects in a Chemical Reaction (*Continued*)

Assuming the specific heat capacity of the solution to be 4.185 MJ/ton K (4.185 J/g K),[11] the temperature rise would be 4380/(4.185 × 6) = 174 K (°C), causing the solution temperature to rise to ~200°C. This is physically impossible, as water will start boiling off from the solution at 100°C at atmospheric pressure. So, in reality, the solution temperature will increase until it reaches 100°C (more accurately, the boiling point of the solution, which would be somewhat higher due to the colligative properties),[12] when boil-off of water will commence. The amount of water boiling off can be calculated as follows:

4380 = Heat used for raising solution temperature from 25°C to 100°C + Heat used for water boil-off

$$4380 = 6 \times 4.185 \times (100 - 25) + m_{steam} \times \Delta H_{vap}$$

Here, m_{steam} is the mass of the steam generated, and ΔH_{vap} is the latent heat of vaporization (2260 J/g or MJ/ton). Therefore, the amount of steam generated, or water boil-off, is 1.105 ton or 1105 kg. The complete material and energy balances for the system are shown in Figure 7.8.

As can be seen from the figure, slaking the lime yields two products: 1105 kg steam and $Ca(OH)_2$ containing ~19% moisture.

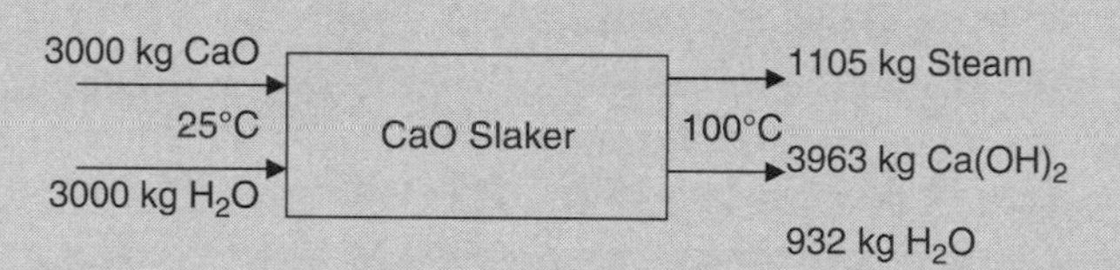

Figure 7.8 Material and energy balance on a CaO slaking operation.

7.3 Summary

The principle of the conservation of energy forms the basis of energy balance computations. This chapter presented the different forms of energy of interest in chemical processes and the quantification of the principle of conservation of energy in terms of these forms of energy. The concept of enthalpy and its dependence on temperature was also discussed. The application of these

11. A gross simplification, as this value is the specific heat capacity of water, and the system consists of a highly concentrated solution.
12. Colligative properties are solution properties that depend on the amount of solute present in the solution.

principles was then demonstrated, with different problems dealing with quantification of enthalpy changes in transformations and determination of system temperatures. The nature of computational problems ranged from simple linear equations to more complex transcendental equations. The approach to solving these problems and the resulting solutions using various numerical and graphical techniques available from different software programs were also presented.

References

1. Felder, R. M., and R. W. Rousseau, *Elementary Principles of Chemical Process*, Third Edition, John Wiley and Sons, New York, 2005.
2. Reklaitis, G. V., *Introduction to Material and Energy Balances*, John Wiley and Sons, New York, 1983.
3. Smith, J. M, H. C. Van Ness, and M. M. Abbott, *Introduction to Chemical Engineering Thermodynamics*, Seventh Edition, McGraw-Hill, New York, 2005.
4. Kyle, B. G., *Chemical and Process Thermodynamics*, Third Edition, Prentice Hall PTR, Upper Saddle River, New Jersey, 1999.
5. Koretsky, M. D., *Engineering and Chemical Thermodynamics*, Second Edition, John Wiley and Sons, New York, 2012.
6. Prausnitz, J. M., R. M. Lichtenthaler, and E. G. de Azevedo, *Molecular Thermodynamics of Fluid-Phase Equilibria*, Third Edition, Prentice Hall PTR, Upper Saddle River, New Jersey, 1999.

Problems

7.1 What is the thermal duty requirement for heating the *dry air* from 30°C to 60°C before it is fed into the dryer in problem 6.1? The molar density of air is 1.17 mol/ft^3.

7.2 What is the adiabatic flame temperature when 15% excess air is supplied to the natural gas burner in Example 7.2.1? What is the temperature when stoichiometric quantity of oxygen is used for combustion rather than air?

7.3 The CO generated during elemental phosphorus production (problem 6.3) is burned to obtain energy. How much heat can be obtained by burning the CO per ton of phosphorus produced? Enthalpy of formation of CO is –110.5 kJ/mol.

7.4 Dissolution of urea (NH_2CONH_2) in water can be represented by the following equation:

$$NH_2CONH_2 + 50H_2O \rightarrow NH_2CONH_2\ (H_2O)_{50}$$

The species on the right of the equation simply represents dissolved urea. The enthalpies of formation in kJ/mol of the species are urea: –333.0, water: –285.8, dissolved urea: –14608.

What is the enthalpy change in the process (this change is called enthalpy [or heat] of solution)? Assuming that the process is adiabatic, what is the change in temperature of the solution? The specific heat capacity of the solution can be assumed to equal to that of water: 4.185 J/g K.

7.5 The dry air in problem 7.1 is heated from 30°C to 60°C using steam at atmospheric pressure. The temperature of the condensing steam is 100°C. What is the heat exchange area requirement if the overall heat transfer coefficient is 150 W/m^2 K?

7.6 The standard enthalpies of formation of ethanol, acetaldehyde, and hydrogen in kJ/mol are –235.3, –166, and 0, respectively. What is the energy change for the dehydrogenation reaction described in problem 6.10? Is the reaction endothermic or exothermic? The reaction is conducted at 545 K, and the product gases are cooled to 360 K. What is the heat removal rate for a continuous reactor fed with 1 mol/s of pure ethanol? Assume that the ethanol conversion is 35%. The heat capacities of the three species in J/mol K are $C_{P,ethanol}$: 87.5; $C_{P,acetaldehyde}$: 55.2; $C_{P,hydrogen}$: 29.

7.7 How will the heat duty change if the reactor feed to the dehydrogenation reaction described in problem 7.6 consists of an equimolar stream of ethanol and nitrogen? Assume that other conditions—ethanol feed rate, temperature, conversion—are the same as described in problem 7.6. The heat capacity of nitrogen in J/mol K is described by the following equation, where T is the temperature in K:

$$C_{P,nitrogen} = 29 + 1.85 \cdot 10^{-3}\, T - 9.65 \cdot 10^{-6}\, T^2 + 16.64 \cdot 10^{-9}\, T^3 + 117/T^2$$

7.8 The heat capacity of hydrogen gas was measured at various temperatures, and the data follows:

T, K	C_P, J/g K	T, K	C_P, J/g K	T, K	C_P, J/g K
175	13.12	850	14.77	2100	17.18
200	13.53	900	14.83	2200	17.35
225	13.83	950	14.90	2300	17.50
250	14.05	1000	14.98	2400	17.65
275	14.20	1050	15.06	2500	17.80
300	14.31	1100	15.15	2600	17.93
325	14.38	1150	15.25	2700	18.06
350	14.43	1200	15.34	2800	18.17
375	14.46	1250	15.44	2900	18.28
400	14.48	1300	15.54	3000	18.39
450	14.50	1350	15.65	3500	18.91
500	14.51	1400	15.77	4000	19.39
550	14.53	1500	16.02	4500	19.83
600	14.55	1600	16.23	5000	20.23
650	14.57	1700	16.44	5500	20.61
700	14.60	1800	16.64	6000	20.96
750	14.65	1900	16.83	—	—
800	14.71	2000	17.01	—	—

Obtain an equation (or equations) expressing the heat capacity as a function of temperature. The possible forms of the equation have been described in the chapter. (Hint: First draw a temperature-heat capacity plot. Identify different regions if so indicated by the data trend. It is possible that the heat capacity is described best by using different equations for different regions.)

7.9 Latent heats of vaporization of acetaldehyde and ethanol are 27.7 kJ/mol and 39.0 kJ/mol, respectively. How much energy needs to be removed just for the condensation of the two components from the exhaust gas of problem 7.7?

7.10 The exhaust gases of problem 7.7 are passed through a chiller where they are cooled from 360 K to 293 K. The cooling is affected by a chilled water stream, which is maintained at constant temperature of 5°C. Calculate the heat transfer area needed for condensation only if the heat transfer coefficient is 8 W/m^2 K.

CHAPTER 8

Computations in Chemical Engineering Thermodynamics

The laws of thermodynamics . . . express the approximate and probable behavior of systems of a great number of particles.

—J. Willard Gibbs[1]

The principle of conservation of energy, discussed in the previous chapter, merely states that the total energy of the universe is constant, and interconversions between different forms of energy are exactly balanced. The principle does not offer any indication of the feasibility of a particular energy transformation. No inference can be drawn regarding the spontaneity of the transformation that a system may undergo. Thermodynamics is that branch of physics and engineering science which allows us to determine and quantify the behavior of systems in such interconversions [1]. The principle of conservation of energy appears in thermodynamics as its first law. The second law of thermodynamics provides the basis for determining the direction of the energy transformations that occur spontaneously [2]. The mathematical treatment based on theoretical principles of thermodynamics allows us to determine not only the direction of the transformation but also the efficiency of transformation as well as the final conditions at the end of the transformation. Thermodynamics also allows us to determine the energy requirements for any desired transformation.

The development of classical thermodynamics as a discipline resulted from the study of heat engines—machines that harness thermal energy for performing mechanical work.[2] Chemical engineering thermodynamics involves application of thermodynamic principles for the analysis and

1. Arguably the greatest scientist America has produced, and one whose contributions led to formalization of theoretical principles of thermodynamics and statistical mechanics. Quotation source: Gibbs, J. W., *Elementary Principles in Statistical Mechanics*, Yale University Press, New Haven, Connecticut, 1902.

2. The term *thermodynamics* itself is a combination of two Greek words that mean heat and motion.

prediction of behavior of chemical systems. Chemical engineers employ the principles of thermodynamics to address broadly two types of problems [3, 4]:

- Computations of energy-work interconversions, including determination of maximum amount of work that can be obtained from heat/energy input or minimum amount of work input required to effect a change, and
- Determination of equilibrium state of the system in terms of measurable variables such as temperature, pressure, compositions, and so on.

A simplified discussion of the fundamental concepts of thermodynamics is presented in this chapter followed by some of the basic computational problems often encountered by a chemical engineer.

8.1 Fundamental Concepts of Thermodynamics

The formal concepts of thermodynamics are subtle and require much thought before they are comprehended fully or adequately [2]. The development of these concepts is based on a strong foundation in chemistry, physics, and mathematics, built over several semesters of study. Such development is not attempted here; rather, some essential thermodynamic quantities are introduced in terms of their significances and dependences on measurable properties.

8.1.1 System Definition, Properties, and State

A *system* is any part of the universe under consideration or that is the focus of thermodynamic analysis. For example, the dashed line shown in Figure 6.10 represents the boundary of a system. The system itself is composed of all the units enclosed within the boundary. The part of the universe that is excluded from the system or is outside the system boundary is termed as the *surroundings* [5]. A system may or may not exchange mass and energy with the surroundings; an *open* system is one where both mass and energy are exchanged with the surroundings, whereas a *closed* system, by definition, does not exchange any mass but is able to exchange energy with the surroundings. A system that exchanges neither mass nor energy with the surroundings is termed an *isolated* system [5].

It is clear that each system will exhibit certain characteristics that distinguish it from other systems and surroundings. For example, a system will have certain temperature, pressure, volume, as well as constituents present in certain proportions. The characteristics or *properties* allow us to quantify the system. Certain properties of the system, such as the temperature or pressure, do not depend on the size of the system and are termed *intensive* properties, as stated in the previous chapter. On the other hand, properties such as mass or volume are dependent on the size of the system and are called *extensive* properties [4]. Extensive properties are additive in nature; the magnitude of an extensive property is the sum of the magnitudes of that property for various parts of the system [5].

A *state* of a system refers simply to the conditions present in the system, such that the properties are invariant with respect to time [4]. For a pure component present in a single phase, the state of the system can be specified simply by stating its pressure and temperature. This fixes all of its intensive properties, whereas the extensive properties require specification of the quantity present. In other words, specifying the pressure and temperature of that pure substance automatically fixes the *specific volume* (volume per mole) of the substance. The total volume of the system can be determined if the total number of moles is specified.

The relationship among the pressure, volume, and temperature describes the *volumetric behavior* for a substance. This behavior, or the *volumetric* properties of the substance, is used in thermodynamics to determine its *thermodynamic* properties [6]. The following sections describe some of these thermodynamic properties.

8.1.2 Internal Energy and Entropy

Following is the mathematical statement of the first law of thermodynamics for a system, neglecting the changes in the mechanical (kinetic and potential) energy:

$$\Delta U = Q - W \tag{8.1}$$

In this equation, Q is the heat added to the system; W, the work done by the system; and ΔU, the change in the internal energy.

The internal energy U can be visualized as the kinetic and potential energy of a substance or system at the atomic and molecular level [7], and is one of the fundamental thermodynamic properties. Significance of the internal energy is linked to the first law of thermodynamics. If no heat is supplied (or removed) from the system, then $-\Delta U = W$; that is, there is a decrease in the internal energy manifests as the work done by the system. Conversely, if work is done on the system, and no heat is supplied or removed from it, the net result is an increase in the internal energy of the system.

As mentioned previously, the first law does not indicate the feasibility of extracting the work from the system. Another fundamental thermodynamic quantity, *entropy*, is needed for this purpose. Entropy, as defined and visualized by Clausius,[3] is a measure of the *transformation content* (capacity to transform) of the system [2]. It was shown by Clausius that only those transformations in which a system experiences a loss in its capacity to transform can occur naturally or spontaneously. Mathematically, *change in entropy is always positive in natural processes*:

$$[dS]_{adiabatic} \geq 0 \tag{8.2}$$

Here, S is the entropy. The subscript *adiabatic* indicates that the process occurs in absence of any heat (energy) exchange between the system and the surroundings. This is one of the most useful formulations of the second law of thermodynamics [5]. The second law is often presented with several alternative formulations, all of which are equivalent. The details of these formulations are beyond the scope of this book and are generally discussed in the engineering thermodynamics and chemical engineering thermodynamics courses.

The utility of the concepts of internal energy and entropy in devising solutions to the first type of problems described should be apparent at this point. If the change in the internal energy can be determined for a system, then the amount of work that can be obtained from the system can be calculated. The changes occurring in the system (and surroundings, if needed) can also be determined. The calculation of change in the entropy allows us to determine whether or not the proposed transformation of the system is feasible naturally.

8.1.3 Enthalpy and Free Energies

As interconversion between heat and work is the central concern of thermodynamics, it is useful to define thermodynamic properties that are related to

---

3. Rudolf Clausius, 19th-century German mathematician and scientist, was a major contributor to and one of the founders of the discipline of thermodynamics.

the heat content of a system and energy available for conversion to work. This consideration leads to three thermodynamic functions or properties that are mathematically defined as follows:

$$\text{Enthalpy } (H)\text{: } H = U + PV \tag{8.3}$$

$$\text{Helmholtz energy } (A)\text{: } A = U - TS \tag{8.4}$$

$$\text{Gibbs energy}^4 \text{ } (G)\text{: } G = H - TS \tag{8.5}$$

In these equations, P, V, and T are the pressure, volume, and temperature of the system.

Enthalpy of a system is a measure of its heat content, which was discussed in Chapter 7, "Energy Balance Computations." The significances of the two thermodynamic properties are best understood in terms of changes in the values of A and G. Simply stated, the change in Helmholtz energy represents the maximum amount of work that can be obtained from the system, whereas the change in Gibbs energy represents the maximum amount of work that can be extracted from the system *excluding any work of expansion*. Mathematically [6], this is as follows:

$$\Delta A = -W_{\text{max}} \tag{8.6}$$

$$\Delta G = -\left(W_{\text{max}} - P\Delta V\right) \tag{8.7}$$

Here, W_{max} represents the maximum work that can be extracted from the system. Equations 8.6 and 8.7 indicate that any work extracted from the system (done by the system) is at the expense of the Helmholtz energy and Gibbs energy, respectively. Gibbs energy is also the key thermodynamic property in determination of system equilibrium, as will be discussed later.

8.1.4 Property Changes in Transformations

A system's properties change when it moves from one state to another. This transformation or process of changing state is inevitably accompanied by heat and work exchange as governed by the first law. When the work and

4. The terms *Helmholtz function* (or Helmholtz energy) and *Gibbs function* (or Gibbs energy) are preferred over classical usage of terms *Helmholtz free energy* and *Gibbs free energy*. Helmholtz function is named after Herrmann Helmholtz, another central founder of the discipline of thermodynamics.

heat effects associated with the process are such that it is feasible to restore the system to its original state, then the process is *reversible*. Effectively, if it is possible to move the system from its initial state to a final state, then reverse the process ending up at the initial state while restoring both the system and the surroundings exactly to their original conditions, then both the forward and reverse steps are considered to be reversible [4]. However, real processes are invariably accompanied by some dissipative phenomena, such as friction, with the result that it is never possible to restore both system and the surroundings to their original conditions. Real processes are thus *irreversible* by nature. The concept of reversibility and reversible process is, however, extremely important in thermodynamics, as a reversible process represents a limiting condition that provides a bound for a real or actual process. Changes occurring in a reversible process can be computed from the thermodynamic properties and used to estimate the actual changes that can be expected in a real process.

Certain system properties and changes in these properties can be determined from the knowledge of the initial and final states of the systems. Such properties or functions do not require any information about the actual process of transformation of the system. Such properties are variously referred to as *state properties*, *state variables*, *state functions*, and so on [4]. Temperature, pressure, entropy, and internal energy are some examples of a state property.

Conversely, some of the properties or variables depend not only on the initial and final states of the system but also on the path traversed by the system while undergoing the transformation. Such properties are called *path functions*, *path variables*, and so on. Work done by the system or on the system is an example of a path function. Thermodynamic processes are often represented on a pressure-volume (P-V) diagram, which shows the system pressure as a function of its volume. System work is represented by the area under the curve representing the system transformation on the P-V diagram, as shown in Figure 8.1 [5].

Figure 8.2 shows two alternative paths followed by the system while undergoing transformation from state 1 to state 2. It can be clearly seen that the areas under the curve bound by volumes represented by points 4 and 6 are different when a direct path is followed from 1 to 2, as opposed to the path followed by first going from 1 to 3 and then from 3 to 2. As previously mentioned, these areas represent the system work, which is a path function. Other thermodynamic property changes (ΔU, ΔS, etc.) do not depend on which path is followed between 1 and 2. They are state functions and depend only on the initial and final states.

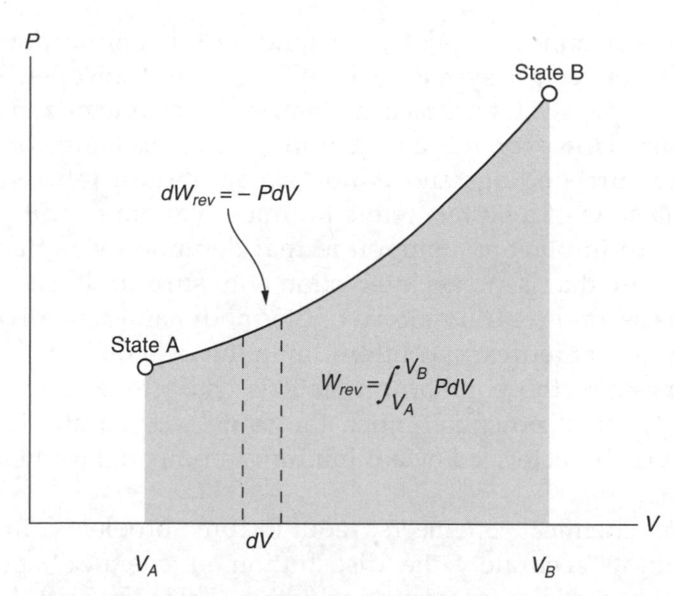

Figure 8.1 Representation of work done in a reversible process on a *P-V* diagram.
Source: Matsoukas, T., *Fundamentals of Chemical Engineering Thermodynamics with Applications to Chemical Processes*, Prentice Hall, Upper Saddle River, New Jersey, 2013.

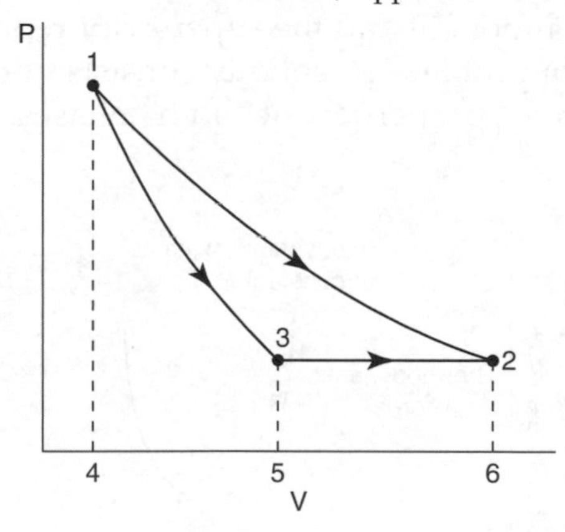

Figure 8.2 Work, a path-dependent function, as illustrated by areas under the curves 1-2 and 1-3-2.
Source: Kyle, B. G., *Chemical and Process Thermodynamics*, Third Edition, Prentice Hall, Upper Saddle River, New Jersey, 1999.

8.1.5 Chemical Potential and Equilibrium

All naturally occurring processes proceed spontaneously until the state of *equilibrium* is reached where no further net change occurs in the system. The implication of the equilibrium conditions is that the system is not interacting with the surroundings [4]. Understanding this implication is crucial to distinguish between a system at equilibrium and an open system at steady state. The open system, at steady state, is also characterized by time-invariant properties. However, it is engaged in the interexchange of mass and energy with the surroundings and is not at equilibrium [4]. As stated in section 8.1.1, the state of a system refers to time-invariant conditions present in the system. An implicit assumption in that definition was that the system is at equilibrium; that is, no net interaction with surroundings is occurring.

The second central concern of thermodynamics involves identifying the conditions that represent equilibrium in the system. In a chemical thermodynamic system, the equilibrium state is characterized by the minimum in thermodynamic potential, much the same way equilibrium in mechanical systems is characterized by a minimum in potential (or height), as shown in Figure 8.3.

For chemical engineers, equilibrium problems invariably involve determining accurately the distribution of chemical species in different phases that are in contact with each other. This phase equilibrium problem is shown in Figure 8.4 [6].

Phases α and β, each consisting of the same N components, coexist in contact with each other at a pressure P and temperature T. The composition in each phase is represented by the mole fractions of the components, the subscript representing the component and the superscript representing the phase. The phase equilibrium problem essentially consists of complete characterization of the intensive properties of both phases. In this

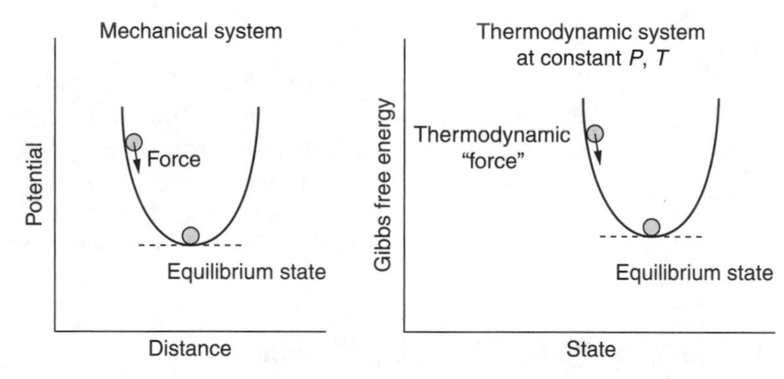

Figure 8.3 Minimization principle in mechanical and thermodynamic systems.

Source: Matsoukas, T., *Fundamentals of Chemical Engineering Thermodynamics with Applications to Chemical Processes*, Prentice Hall, Upper Saddle River, New Jersey, 2013.

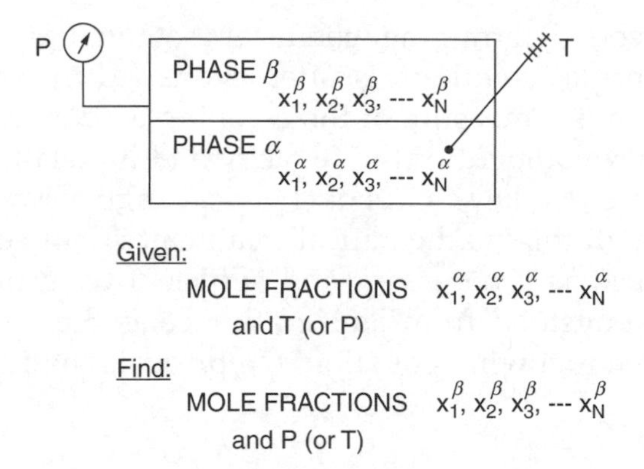

Figure 8.4 Essence of a phase equilibrium problem.
Source: Prausnitz, J. M., R. M. Lichtenthaler, and E. G. de Azevedo, *Molecular Thermodynamics of Fluid-Phase Equilibria*, Third Edition, Prentice Hall, Upper Saddle River, New Jersey, 1999.

particular problem, the composition of phase α is known along with either the temperature (or the pressure). The chemical engineer is required to find the composition of the other phase β and the pressure (or the temperature).

The concept of *chemical potential* provides the foundation for determining these equilibrium conditions. It can be readily understood that when a species is present in two phases that are in contact with each other and is able to distribute itself in both the phases—that is, it is able to cross the phase boundary that separates the two phases—then it will continue to do so until no driving force exists for its movement across the phase boundary. This driving force for the movement of the species across the phases is provided by the difference in the chemical potentials of the species in the two phases, much the same way the temperature difference between two bodies in contact provides the driving force for heat transfer. It follows that this two-phase system will reach equilibrium and there will be no net movement of the species between the phases when its chemical potentials in the two phases are equal, again similar to absence of heat transfer between two bodies that are present at identical temperatures. Mathematically, the equilibrium condition is represented as follows [6]:

$$\mu_i^\alpha = \mu_i^\beta \tag{8.8}$$

Here, μ represents the chemical potential,[5] subscript i refers to the species i, and α and β are the two phases.

Equation 8.8 provides the basis for determining the equilibrium state for the system: It requires computing the chemical potentials of the species

5. Chemical potential of a species is also equal to its *partial molar Gibbs energy*, the discussion of equivalence left to the courses in thermodynamics.

in the two phases. However, determining chemical potential is rather complicated, and the equilibrium condition is often expressed in terms of quantity called *fugacity*, which is a measure of the escaping tendency of the species. The mathematical development of the fugacity and its relationship with the chemical potential are two key topics of chemical engineering thermodynamics courses. Although this mathematical treatment is not covered here, the equality of fugacities as a necessary and sufficient condition for equilibrium can be easily understood from its significance as the escaping tendency of the species. In the following equation, f_i represents the fugacity of the species i:

$$f_i^{\alpha} = f_i^{\beta} \tag{8.9}$$

where f_i^{α} is the escaping tendency of species i from the phase α. Since the phase α is in contact with the phase β, the species i will escape into the phase β. However, the species also has a fugacity f_i^{β} in phase β, which indicates its tendency to escape from phase β to phase α. So long as the two fugacities are not equal, the species will move from one phase to another depending on in which phase it has a higher fugacity. However, once the two fugacities are equal, equilibrium is reached with no net transfer of the component, as there is no higher preference for escaping from either phase.

The determination of equilibrium of a system is thus essentially a matter of computing the fugacities of the species that are present in the phases in contact with each other. Fugacities, as with other thermodynamic properties, can be readily calculated from the volumetric properties of substances. The accuracy of fugacities is critically dependent on having an accurate mathematical expression that can explain the relationship among the pressure, temperature, and volume of a substance.

Equilibrium in reacting systems is based on similar principles. The equilibrium constant for a reaction can be related to the fugacities and, in turn, to the concentrations/pressures of the species involved in the reaction [8]. The equilibrium constant can be determined from the thermodynamic properties, and concentrations of species or conversion of a reaction can be determined from the equilibrium constant. Analyses of multiphase reacting systems involve simultaneous application of phase and reaction equilibria.

8.1.6 Volumetric Behavior of Substances

The thermodynamic quantities internal energy, entropy, enthalpy, Helmholtz energy, Gibbs energy, and chemical potential provide the framework for the solution of both types of problems previously described. Although the absolute values of the thermodynamic quantities cannot be determined, *changes in these properties* can be computed accurately. However, these computations require the knowledge of the volumetric behavior of the substance; that is, the quantitative relationship between the pressure, volume and temperature for the substance. Figure 8.5 shows a typical representation of the volumetric behavior of a pure substance on a P-V diagram [4]. The solid lines are *isotherms*, representing the relationship between the pressure and volume at a constant temperature.

Point C in the figure represents the critical point, with coordinates of V_C (critical volume) and P_C (critical pressure). The isotherm at the critical temperature T_C is tangent to the curve represented by the dashed line. This curve represents a boundary between the two-phase vapor-liquid region and the single-phase region. Isotherms at supercritical temperatures ($T > T_C$) lie entirely within the gas phase region. The volumetric behavior is more complicated at subcritical temperatures, where depending on the pressure, both liquid and vapor coexist.

The volumetric behavior of a pure substance can also be represented on a pressure-temperature (P-T) diagram, as shown in Figure 8.6 [5]. Point F represents the triple point for the substance where all three phases—solid, liquid, and vapor—coexist. The solid lines from F represent the boundaries between different phases, F-S represents the boundary between solid and liquid, and F-C represents the boundary between liquid and vapor. C, as in Figure 8.5, is the critical point, and the supercritical region has pressure and

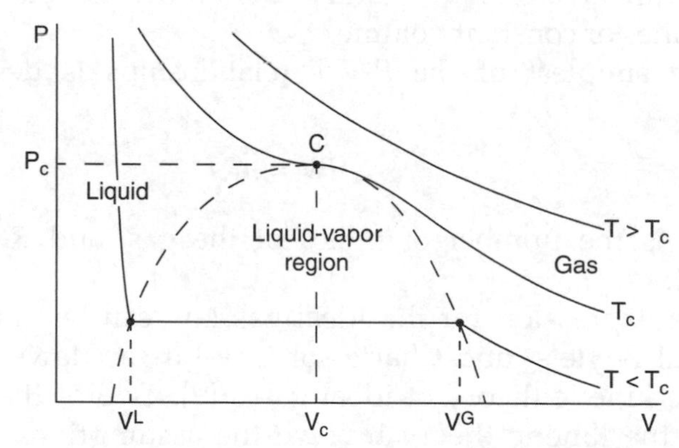

Figure 8.5 Pressure-volume phase diagram for a pure substance.

Source: Kyle, B. G., *Chemical and Process Thermodynamics*, Third Edition, Prentice Hall, Upper Saddle River, New Jersey, 1999.

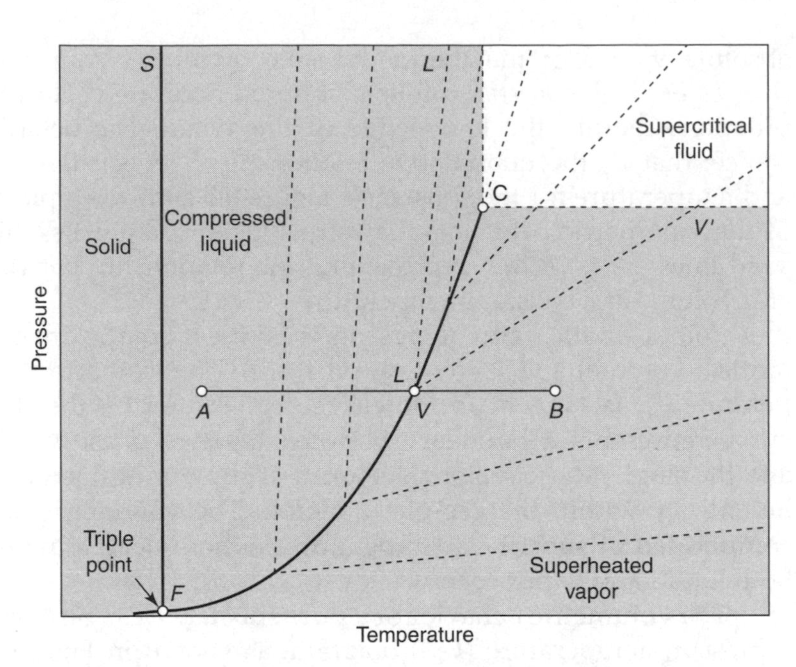

Figure 8.6 Pressure-temperature phase diagram for a pure substance.

Source: Matsoukas, T., *Fundamentals of Chemical Engineering Thermodynamics with Applications to Chemical Processes*, Prentice Hall, Upper Saddle River, New Jersey, 2013.

temperature greater than P_C and T_C, as shown. The dashed lines are *isochores*, that is, lines of constant volume.

The simplest of the *P-V-T* relationships is described by the ideal gas law:

$$PV = nRT \tag{8.10}$$

Here, n is the number of moles of the gas, and R is the universal gas constant.

The expression for the ideal gas law can be obtained by combining empirical Boyle's[6] and Charles or Gay-Lussac[7] laws, or it can be derived from the kinetic theory of ideal gases [9]. Two of the fundamental postulates of the kinetic theory involve the assumptions of no intermolecular

6. Gas volume is inversely proportional to its pressure.
7. Gas volume is directly proportional to its absolute temperature.

forces and negligible volume of molecules. However, molecules of real substances do occupy certain volume, small though it may be. Further, these molecules do exert different types of attractive and repulsive forces on each other, with the result that the behavior of real substances cannot be accurately described by the ideal gas law. It follows that this lack of accuracy would lead to errors in the computation of thermodynamic properties. An accurate mathematical description of the volumetric behavior of substances—the equation of state (EOS)—is a fundamental need for any thermodynamic calculations.

Several different types of EOSs have been reported in literature [5]. Some of the equations are empirical, some are semi-empirical, and many others are developed from first principles. One of the earliest equations describing the behavior of real gases is the van der Waals equation, written in its pressure-explicit form [5]:

$$P = \frac{RT}{v-b} - \frac{a}{v^2} \tag{8.11}$$

In this equation, v is the molar volume of the gas ($=V/n$), and a and b are the characteristic constants for the gas. The constant a is related to the attractive force between the molecules, and b is the effective physical volume occupied by a mole of the gas. These two constants are typically calculated from two other characteristic constants unique to each substance: the critical pressure P_c and critical temperature T_c. The critical constants for a large number of substances are available from the sources mentioned in Chapter 7.

The van der Waals equation belongs to a class of EOSs called the cubic equations of state, as they form a third-order polynomial in v. Several other types of EOSs are available that may describe the volumetric behavior of some classes of compounds accurately but fail for others. A chemical engineer must select the appropriate EOS for the system under consideration and perform computations using that equation.

8.1.7 Nonideality

As mentioned in the previous section, the volumetric behavior of a real substance rarely conforms to the ideal gas law. This nonideal behavior requires developing an accurate EOS for determining thermodynamic property changes in processes.

Most of the process streams in practice typically consist of mixtures. Even pure product streams typically contain (tolerable) levels of impurities. The complexity of behavior of mixtures and resultant deviation from the idealized behavior increases with the increasing complexity of interactions among the mixture constituents. The extensive properties are rarely additive; for example, the total volume of a liquid mixture is invariably less than the sum of the individual volumes mixed together. It follows that the intensive properties will also differ significantly from what can be estimated from mole- or mass-fraction weighted individual intensive properties. This results in nonideal behavior of the system, which has a significant impact both on energy-work interconversions and equilibrium conditions. The mathematical treatment of nonideality, particularly with respect to equilibrium problems, involves refining fugacity computations through determination of fugacity coefficients (φ_i) or activity coefficients (γ_i). This complex treatment is discussed in chemical engineering thermodynamics courses.

The next section provides an introduction to some of the computational problems in chemical engineering thermodynamics.

8.2 Basic Computational Problems

Figure 8.7 shows a process on the P-V diagram with an arbitrary process path, wherein a system undergoes a change from its initial state to a final state along the path shown by the solid line.

As stated earlier in the chapter, the types of computational problems, at the elementary level, involve estimations of the changes in the thermodynamic properties. For those properties that are state functions, the approach to solving these problems and determining the changes in the thermodynamic quantities involves postulating an alternative path from the initial state of the system to its final state, as shown by the dashed lines in Figure 8.7. The system first undergoes a constant volume process at $V_{initial}$, labeled 1, followed by a constant pressure process at P_{final}, labeled 2. Computations of thermodynamic property changes along these constant volume and constant pressure paths present a much more manageable challenge as compared to those along the actual path followed by the process. The generalized principle of this approach is shown Figure 8.8 [6]. Appropriately designed, it is possible to determine the changes in the required quantities in each of the segments constituting the postulated alternate path. The overall change in the

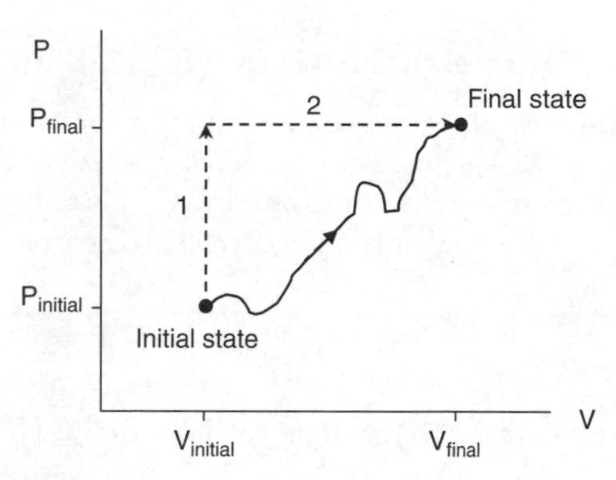

Figure 8.7 An arbitrary thermodynamic process.

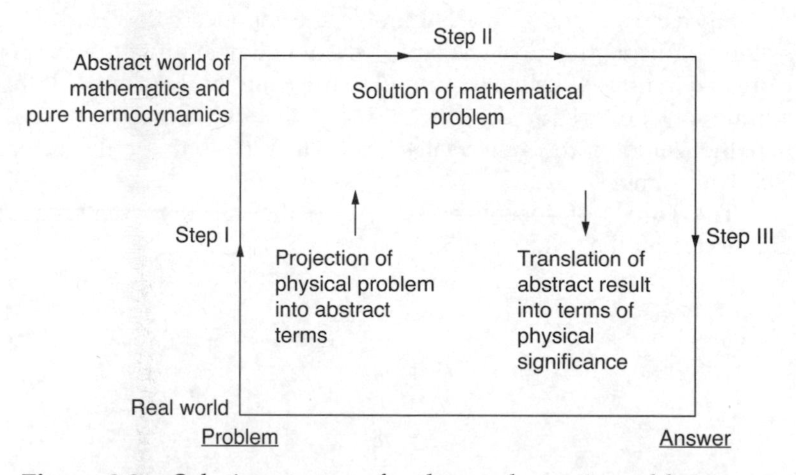

Figure 8.8 Solution strategy for thermodynamic problems—path formulation, abstraction, and interpretation of results.
Source: Prausnitz, J. M., R. M. Lichtenthaler, and E. G. de Azevedo, *Molecular Thermodynamics of Fluid-Phase Equilibria*, Third Edition, Prentice Hall, Upper Saddle River, New Jersey, 1999.

transformation of the system from the initial to final state is simply the sum of the changes in these individual segments.

Next, as mentioned earlier, the problem of determining equilibrium involves estimating the fugacities of the components. Appropriate mathematical expressions are available for various situations to determine the fugacities from the volumetric properties of the substances. The following examples provide an idea of the computational problems in chemical engineering thermodynamics.

Example 8.2.1 Volumetric Flow Rate of Methane

Calculate the volumetric flow rate of methane to the natural gas burner described in problem 7.2.1 using both the ideal gas law and van der Waals EOS. Assume that the pressure is atmospheric. What is the error in using the ideal gas law if the van der Waals EOS correctly describes the volumetric behavior of methane?

Solution

The molar volume, according to the ideal gas law, follows:

$$v_{ideal} = RT/P = 0.082 \times 298.15/1$$

$$= 24.436 \text{ L/mol}$$

Since the molar flow rate is 1 mol/s, the volumetric flow rate of 24.436 L/s is obtained using the ideal gas law. It should be noted that the temperature is expressed in units of kelvin (K) and pressure in atmosphere (atm). If the specific volume of the gas is desired in the units of liters per mole (L/mol), then the value of the universal gas constant R used for the calculation is 0.082 L atm/mol K.

The values of constants a and b in the van der Waals equation are obtained from the following [5]:

$$a = \frac{27}{64}\frac{R^2T_C^2}{P_C} \tag{E8.1}$$

$$b = \frac{RT_C}{8P_C} \tag{E8.2}$$

The critical temperature and pressure for methane are 190.6 K and 45.6 atm, respectively.

The values of a and b calculated from these data and the resulting van der Waals EOS at the given pressure and temperature are as follows:

$$a = 2.29 \text{ L}^2 \text{ atm/mol}^2 \qquad b = 0.0428 \text{ L/mol}$$

$$1 = \frac{0.082 \cdot 298.15}{v_{vdw} - 0.0428} - \frac{2.29}{v_{vdw}^2} \tag{E8.3}$$

Equation E8.3 is a cubic equation in v_{vdw}, which can be solved using techniques described in the earlier chapters. Using the Goal Seek function in Excel yields a molar volume value of 24.387 L/mol (the value obtained from the ideal gas law was used as the initial guess for obtaining the solution).[8] The volumetric flow rate is then equal to 24.387 L/s. The percentage of error if the ideal gas volume is used is 0.2%, which is quite low under these conditions.

It is instructive to see how the error changes when the conditions are changed. If the pressure is changed to 200 atm, then the ideal gas molar volume is 0.122 L/mol, yielding a volumetric flow rate of 122 mL/s. The molar volume, according to the van der Waals EOS, is computed to be 0.099 L/mol, yielding a volumetric flow rate of 99 mL/s. Assuming that the van der Waals EOS yields a more accurate estimate of the actual volume, an engineer will incur an error of approximately 23% if this calculation is performed assuming ideality of gas behavior. An error of this magnitude is unacceptable in design and operation of the process, reinforcing the need for an EOS that can accurately predict the volumetric behavior of the substances.

Example 8.2.2 Equilibrium Moisture Content of Exhaust Gases

Determine the quantity of water that will condense when the hot exhaust gas of problem 7.2.2 is cooled to 25°C.

Solution

Figure 8.9 is a schematic representation of the physical process occurring as the hot exhaust gases exchange heat with the cooling water and are cooled in turn to 25°C (298 K). Some of the water condenses, resulting in two outlet streams—one gas and the other liquid—exiting the process. Neither CO_2 nor N_2 undergo any condensation at this temperature, and the liquid phase consists only of water. However, some water is present in the gas stream as well, as not all of the water can condense. The component material balance readily yields the moles of water condensing $m = (2 - n)$ mol/s, leaving n as the only unknown in the problem.

(Continues)

8. A cubic equation should have three roots, meaning that there are three values of v that satisfy equation E8.3. However, only one root is positive and real; the other two are complex and are not valid solutions for the molar volume.

Example 8.2.2 Equilibrium Moisture Content of Exhaust Gases (*Continued*)

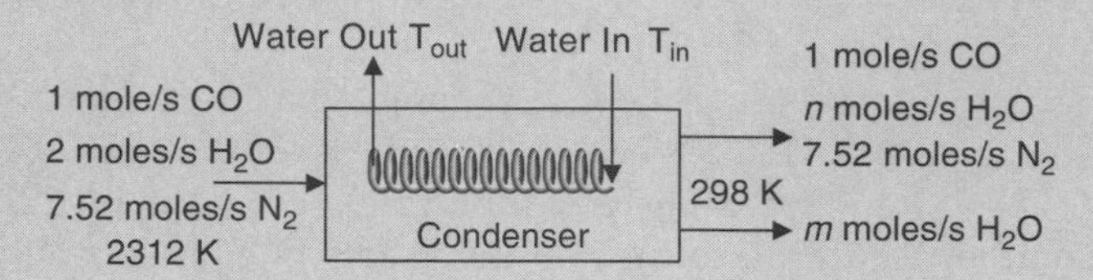

Figure 8.9 Condensation of water vapor.

The concept of equilibrium is used to solve for n. The two outlet streams are in equilibrium with each other, and hence, according to equation 8.9, the fugacity of the water vapor in the gas phase must equal the fugacity of the condensed water. The fugacity of the water vapor is equal to its partial pressure, and the fugacity of the liquid water is equal to the vapor pressure (also called *saturation pressure*) at its temperature.[9] The vapor pressure of water at 25°C (298 K) is 23.8 mm Hg (from thermodynamic data sources), and assuming that outlet streams are exhausted to the atmosphere (pressure 760 mm Hg):

$$\frac{n}{1+n+7.52}760 = 23.8 \quad \text{(E8.4)}$$

The term in front of 760 represents the mole fraction of water vapor in the gas. Solving equation E8.4, we get the following:

$$n = 0.275 \text{ mol/s and } m = 1.725 \text{ mol/s}$$

The condensate flow rate, therefore, is 1.725 × 18 (molecular weight of water) = 31 g/s, or the volumetric flow rate is ~1.86 L/min. The mole fraction of water vapor in the gas phase is found to be 0.031.

Example 8.2.3 Diffusion Equilibrium in a Membrane System

A closed chamber at 40°C is separated into two equal parts by a membrane permeable only to helium. The left part is filled with 0.99 mol of ethane and 0.01 mol helium. The right part has 0.01 mol helium and 0.99 mol nitrogen. What are the mole fractions of helium in each part of the chamber at equilibrium? The final pressures were 37.5 bar in the left part and 84.5 bar in the right part of the chamber.

9. This is the idealization of the system behavior; however, it is an excellent approximation for the system under consideration here. In many other situations, nonideality of mixtures must be accounted for.

Solution

The initial schematic of the system is shown in Figure 8.10.

The solution to this problem requires the application of equation 8.9 to helium:

$$f_{He}^{L} = f_{He}^{R} \tag{E8.5}$$

The superscripts denote the left and right parts of the chamber.

It should be noted that helium is the only component present on both sides of the membrane, and there is no escaping tendency on the part of ethane or nitrogen from the part of the chamber where they exist. The fugacity is expressed in terms of the total pressure of the chamber P, the mole fraction of helium in the chamber (y_{He}), and the fugacity coefficient φ_{He}:

$$f_{He}^{L} = \phi_{He}^{L} y_{He}^{L} P^{L} \tag{E8.6}$$

Equation E8.6 expresses fugacity of helium on the left part of the chamber. A similar equation can be written for its fugacity on the right part of the chamber. The fugacity coefficient is obtained from the EOS for the mixture. When *virial EOSs* (see problem 8.8) are used to describe the volumetric behavior of the mixtures on either side of the chamber, the resulting equation is as follows:

$$\begin{aligned} &\exp\left[\begin{bmatrix} 2\left(y_1^L B_{11} + \left(1-y_1^L\right)B_{12}\right) - \left(y_1^L\right)^2 B_{11} + 2y_1^L\left(1-y_1^L\right)B_{12} \\ +\left(1-y_1^L\right)^2 B_{22} \end{bmatrix} \frac{P^L}{RT}\right] y_1^L P^L \\ &= \exp\left[\begin{bmatrix} 2\left(y_1^R B_{11} + \left(1-y_1^R\right)B_{13}\right) - \left(y_1^R\right)^2 B_{11} + 2y_1^R\left(1-y_1^R\right)B_{13} \\ +\left(1-y_1^R\right)^2 B_{33} \end{bmatrix} \frac{P^R}{RT}\right] y_1^R P^R \end{aligned} \tag{E8.7}$$

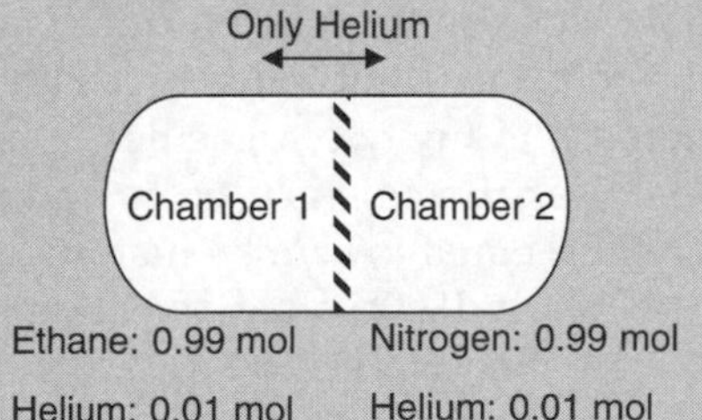

Figure 8.10 Helium transport across a membrane.

(*Continues*)

Example 8.2.3 Diffusion Equilibrium in a Membrane System (*Continued*)

Here B is the second virial coefficient, the two numbers in the subscript of B indicating two components with the numbers 1, 2, and 3 referring to components helium, ethane, and nitrogen, respectively, and the superscript L and R referring to the left and right parts of the chamber. Comparison of equations E8.6 and E8.7 indicates that the exponential terms in the second equation are the fugacity coefficients. The values of various virial coefficients (in cm^3/mol) are $B_{11} = 17.17$, $B_{12} = 24.51$, $B_{22} = -169.23$, $B_{23} = -44.39$, $B_{33} = -1.55$, and $B_{13} = 21.97$.

Equation E8.7 contains two unknowns: the mole fractions of helium on the left and right sides of the chamber. The material balance on helium allows us to express both these mole fractions in terms of moles of helium on one of the sides of the chamber. If there are n_1^L moles of helium on the left side of the chamber at equilibrium, then we get the following:

$$y_1^L = \frac{n_1^L}{0.99 + n_1^L} \tag{E8.8}$$

$$y_1^R = \frac{0.02 - n_1^L}{1.01 - n_1^L} \tag{E8.9}$$

Equations E8.7, E8.8, and E8.9 form a system of three equations in three unknowns that can be solved using any appropriate software.

Solution (using Excel)

The solution using Excel is shown in Figure 8.11.

The values of the pressures of the two parts, the gas constant, temperature, and the virial coefficients are entered in cells A2 through I2. A guess for the value of the moles of helium on the left part of the chamber is entered in cell A8. Cells B8 through F8 contain the values of moles of helium on the right part of the chamber, the mole fractions on the left and right parts of the chamber, and the fugacities on the left and right sides of chambers. An objective function equal to the difference in the two fugacities is entered in cell G8 (=E8-F8). Then the Excel Goal Seek function is used to find the solution by telling it to set the value of cell G8 equal to zero by manipulating cell A8. As can be seen from the figure, nearly 0.005 mole of helium diffuses from the right side to the left side of the membrane. The fugacity of helium at equilibrium is 0.478 bar.

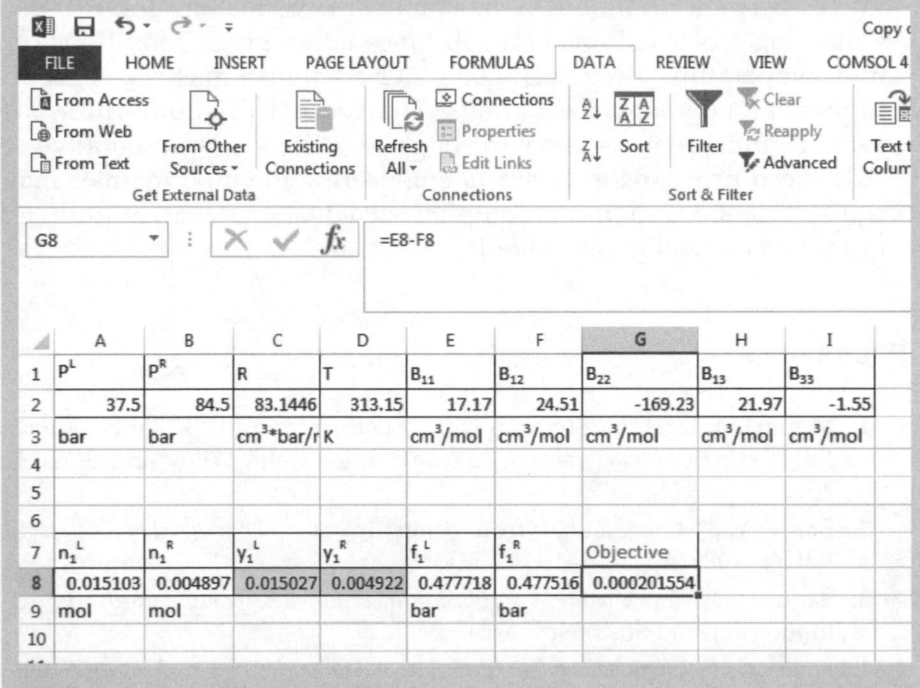

Figure 8.11 Excel solution to example 8.2.3.

The exercise for obtaining the solution to the problem using Mathcad and confirming that the same result is obtained is left to the reader.

These three examples provide an idea of the nature of computational problems in chemical engineering thermodynamics. The student will encounter such problems in the junior year of the study and beyond in his/ her career.

8.3 Summary

Thermodynamics offers a theoretical framework for the quantification of energy-work interconversions and system equilibrium. The fundamental thermodynamic quantities are introduced in this chapter and the significances of these quantities explained. The importance of accuracy in the volumetric behavior of substances is described and illustrated using the van der

Waals equation belonging to the family of EOSs that are cubic in the volume. Obtaining molar volume of a substance under given conditions of pressure and temperature using such cubic EOSs requires mastery of solution techniques for polynomial equations. The concept of equilibrium is also presented and demonstrated through a simple representative example. Advanced problems in chemical engineering thermodynamics may involve transcendental and more complicated equations that require techniques described elsewhere in the book.

References

1. Moran, M. J., H. N. Shapiro, D. D. Boettner, and M. B. Bailey, *Fundamentals of Engineering Thermodynamics*, Eighth Edition, John Wiley and Sons, New York, 2014.
2. Baron, M., "With Clausius from energy to entropy," *Journal of Chemical Education*, Vol. 66, 1989, pp. 1001–1004.
3. Sandler, S. I., *Chemical and Engineering Thermodynamics*, Third Edition, John Wiley and Sons, New York, 1999.
4. Kyle, B. G., *Chemical and Process Thermodynamics*, Third Edition, Prentice Hall, Upper Saddle River, New Jersey, 1999.
5. Matsoukas, T., *Fundamentals of Chemical Engineering Thermodynamics with Applications to Chemical Processes*, Prentice Hall, Upper Saddle River, New Jersey, 2013.
6. Prausnitz, J. M., R. M. Lichtenthaler, and E. G. de Azevedo, *Molecular Thermodynamics of Fluid-Phase Equilibria*, Third Edition, Prentice Hall, Upper Saddle River, New Jersey, 1999.
7. Smith, J. M., H. C. Van Ness, and M. M. Abbott, *Introduction to Chemical Engineering Thermodynamics*, Seventh Edition, McGraw-Hill, New York, 2005.
8. Balzhiser, R. E., M. R. Samuels, and J. D. Eliassen, *Chemical Engineering Thermodynamics: The Study of Energy, Entropy, and Equilibrium*, Prentice Hall, Upper Saddle River, New Jersey, 1972.
9. Maron, S. H., and C. F. Prutton, *Principles of Physical Chemistry*, Fourth Edition, MacMillan Company, New York, 1965.

Problems

8.1 Calculate the volumetric flow rate of methane in Example 8.2.1 using the following Peng-Robinson EOS [5]. Assuming that this EOS gives the most accurate estimate, what is the error in using the ideal gas law? What is the improvement in accuracy over the van der Waals EOS?

$$P = \frac{RT}{v-b} - \frac{a(T)}{v(v+b)+b(v-b))}$$

where

$$a(T) = 0.45724\frac{R^2T_C^2}{P_C}\left(1+k\left(1-\left(\frac{T}{T_C}\right)^{0.5}\right)\right)^2$$

$$b = 0.07780\frac{RT_C}{P_C}$$

$$k = 0.37464 + 1.5422\omega - 0.26922\omega^2$$

ω, the acentric factor for methane, has a value of 0.011.

8.2 Repeat the calculations in problem 8.1 when the pressure is 200 atm.

8.3 Calculate the molar volume of nitric oxide (NO) at 1 bar pressure and 298 K using the ideal gas law, the van der Waals EOS, and the Peng-Robinson EOS. The property data for NO are as follows: $P_c = 64.8$ bar, $T_c = 180$ K, $\omega = 0.588$. The value of the gas constant is 0.08314 L bar/mol K.

8.4 Calculate the volumetric flow rate of the condensate and the mole fraction of water vapor in the gas phase, if the exhaust gases in Example 8.2.2 are cooled to 30°C. The vapor pressure of water at 30°C is 31.8 mm Hg.

8.5 How will the answers to the questions in Example 8.2.2 change when 15% excess air is used for combustion? The material balances in Chapter 6, "Material Balance Computations," can be used as basis for these calculations.

8.6 The vapor pressure–temperature relationship (also the boiling point curve) for a substance is of critical importance in distillations. For many hydrocarbons, the vapor pressure dependence on temperature can be described accurately by the following equation that is valid at temperatures greater than their characteristic divergence temperature (T_d):

$$\log P = A + \frac{B}{T} + \frac{C}{T^2} + D\left(T/T_d - 1\right)^n$$

Here, P is the vapor pressure in millimeters of mercury (mm Hg), and temperatures are expressed in K. The characteristic constants for n-nonane are as follows: A = 6.72015, B = −1188.2 K, C = −186,342 K^2, D = 2.2438, T_d = 503 K and n = 2.50.

A distillation column is to be operated at 10 atm pressure for the separation of a hydrocarbon mixture containing n-nonane. What is the boiling point of n-nonane at this pressure?

8.7 The Bunsen coefficient α—the volume of gas dissolved per unit volume of the liquid when the partial pressure of the gas is 1 atm—for a component in a binary mixture was found to be a function of temperature t (in °C):

$$\alpha(t) = 4.9 \cdot 10^{-2} - 1.335 \cdot 10^{-3} t + 2.759 \cdot 10^{-5} t^2 - 3.235 \cdot 10^{-7} t^3 + 1.649 \cdot 10^{-9} t^4$$

Clearly, α is 0.049 at the freezing point of water. To what temperature must the liquid be heated to reduce the Bunsen coefficient to 0.025?

8.8 Virial EOSs express the compressibility (z) of a gas as a series expansion in volume or pressure. The pressure explicit form (also called the *Leiden* form) of the virial equation is:

$$z = \frac{Pv}{RT} = 1 + \frac{B}{v} + \frac{C}{v^2} + \ldots$$

Here, B and C are the second and third virial coefficients, respectively. Oxidation of metals is frequently carried out under high pressure. Calculate the volumetric flow rate of oxygen fed to a reactor at 50 atm, 298 K, if the molar flow rate is 10 mol/s. The second and third virial coefficients for oxygen at this temperature are $-16.1 \cdot 10^{-6}$ m^3/mol and $1.2 \cdot 10^{-9}$ m^6/mol^2, respectively. What is the percentage error if the ideal gas law is used and the virial equation truncated after the second term is used?

8.9 The osmotic pressure of a solution depends on the concentration of the solute in the solution and is commonly described by the following equation:

$$\frac{\pi}{c} = RT\left(\frac{1}{M} + A_2 c + A_3 c^2\right)$$

Here, π is the osmotic pressure, and c is the concentration of the solute of molecular weight M. The coefficients A_2 and A_3 are called the second and third osmotic virial coefficients, respectively. (Note the similarity with the virial equation of state for gases.) The values of the second and the third virial coefficients for glucose at 298 K are 0.003353 mol cm^3/g^2 and 0.011556 mol cm^6/g^3, respectively. If the observed osmotic pressure is 10 atm, what is the concentration of glucose in the solution?

8.10 The work needed for isothermal compression of a mole of van der Waals gas from a molar volume of v_1 to v_2 is given by the following:

$$W = RT\ln\frac{v_1 - b}{v_2 - b} + a\left(\frac{1}{v_1} - \frac{1}{v_2}\right)$$

2.5 kJ energy is expended as work to compress 1 mole of a van der Waals gas from an initial volume of 1.1 L/mol at 37°C. What is the final volume if the van der Waals constants are a = 1.36 atm L^2/mol^2 and b = 0.0385 L/mol. What would be the final volume if the gas were to behave ideally? The isothermal work of compression for an ideal gas is given by $W_{id} = RT\ln(v_1/v_2)$.

CHAPTER 9

Computations in Chemical Engineering Kinetics

A knowledge of the rate, or time dependence, of chemical change is of critical importance for the successful synthesis of new materials.

—Yuan T. Lee[1]

Thermodynamic analysis of systems, using the principles described in Chapter 8, "Computations in Chemical Engineering Thermodynamics," allows us to determine the *magnitude*, *efficiency*, and *directionality* of the changes occurring in the system [1]. Unfortunately, these analyses do not provide any information about the rate at which these changes will occur, or in other words, the time scale needed to effect these changes. Nor do they provide any information about the mechanism by which these changes occur [2]. It can readily be understood that for an intended change to be economically beneficial, it must occur within a reasonable amount of time. For chemical reactions, the information on time scale associated with a change is provided by chemical kinetics, which describes the rate of change of the species involved in the reaction with time [3]. Chemical engineering kinetics (or chemical reaction engineering) combines this information about the rates of reactions with reactor analysis to develop a quantitative approach for the design and analysis of reactors [4].

Reactor design can broadly be visualized in terms of the determination of the type of reactor (stirred tank or pipeline, batch or continuous), reactor volume, reaction time, and other parameters such as the rate of heat transfer, and so on, needed to carry out a reaction in order to realize economic benefit from the species transformation [5]. A chemical engineer must understand and be able to make quantitative prediction of influence of equipment and operational parameters on the output of the reactor [6]. Optimal reactor design is often the key to obtaining the desired products and ultimately the economics of the process. This chapter first presents some elementary concepts in chemical engineering kinetics, then provides examples of simple problems.

---

1. 1986 Nobel Laureate in Chemistry, Yuan T. Lee is known for his work on chemical reaction dynamics. Quotation source: Nobel Lecture, December 1986, www.nobelprize.org/nobel_prizes/chemistry/laureates/1986/lee-lecture.pdf

9.1 Fundamental Concepts of Chemical Engineering Kinetics

The essence of reactor design is obtaining economically optimum specifications for a reactor for a specified duty. For example, let us assume that the market demand for a certain chemical product (let us denote it by R) is 3000 tons per year. The specified duty then might be defined as a production rate of 10 tons per day (tpd) of that chemical, assuming 300 working days every year. The challenge for the chemical engineer, at a very basic level, is finding the optimum volume of the reactor and reaction time that will yield this specified output of product R, starting from another chemical A, which is the raw material or reactant for the process. This situation is shown in Figure 9.1.

The principles of material balance and information on the stoichiometry of the reaction are used to calculate the material requirements for the process, as discussed in Chapter 6, "Material Balance Computations." This information is necessary, but not sufficient, for reactor design. The approach to reactor design, incorporating the principles of chemical engineering kinetics, is based on the component material balances. The generalized component material balance for any species i, written in terms of molar quantities, is shown by equation 9.1.

$$\frac{dN_i}{dt} = F_{i,in} - F_{i,out} + V \cdot r_i \tag{9.1}$$

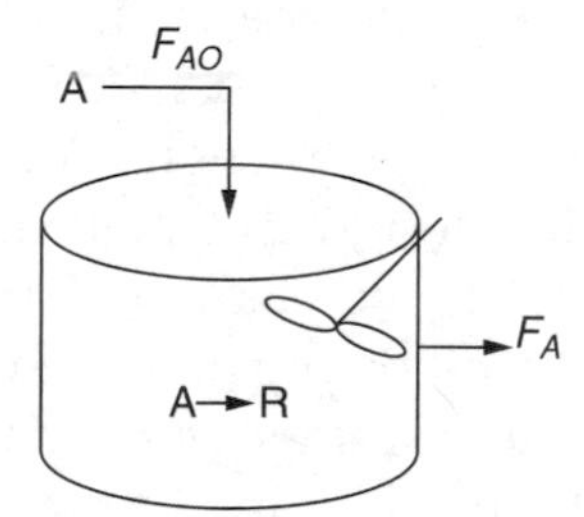

Production rate of R: 10 tpd
Reactor Volume?

Figure 9.1 A simple chemical reactor design problem.
Adapted from: Fogler, H. S., *Elements of Chemical Reaction Engineering*, Fourth Edition, Prentice Hall, Upper Saddle River, New Jersey, 2004.

In this equation, $F_{i,in}$ and $F_{i,out}$ represent the molar flow rates of i entering and leaving the reactor having volume V, respectively. N_i are the moles of the species present in the reactor, and the left side of the equation represents the rate of change of moles of i in the reactor, and r_i is the volumetric rate of generation of moles of i. The last term, which is the product of the reactor volume and the volumetric rate of generation of i, yields the molar rate of generation of i in the reactor. Thus equation 9.1 simply is the mathematical expression stating that the rate at which moles of i changing with time is equal to the difference in the inlet and outlet molar flow rates added to the molar rate of generation of the species i from the reaction.

The *intrinsic kinetics* of the reaction involves quantification of the rate of reaction r_i and its dependence on properties of chemical species involved in the reaction. As mentioned earlier, this intrinsic kinetics is combined with the reactor behavior for reactor design. The basic principles of both these aspects, the intrinsic kinetics and the reactor behavior, constitute the concepts of chemical engineering kinetics and are described in the following section.

9.1.1 Intrinsic Kinetics and Reaction Rate Parameters

The reaction rate for any species i is generally defined as the rate of change of quantity of i (commonly moles of i) per unit time per unit volume of the reactor. Typically, the reaction rate is a function of the concentration of the reacting species and temperature [2]. The mechanism of the reaction, or the exact pathway by which the species reacting transform into reaction products, depends on the nature of the species involved in the reaction. Clearly, the nature of the species also influences the reaction rate. The intrinsic rate of reaction depends only on these factors, and *does not depend on the type of reactor* used for conducting the reaction [4]. It is common to define the rate on the volumetric basis—that is, per unit volume of the reactor—when the reaction is a *homogeneous* reaction, where only a single phase is involved [2]. A large number of chemical reactions are homogeneous, taking place either in the gas or liquid phase. However, many other reactions are *heterogeneous*; that is, they involve two or more phases. Fluid (gas or liquid) phase reactions conducted using solid catalysts are common in the chemical industry. In such cases, the rate may be defined on the basis of the catalyst surface area (rate of change of moles per unit time per unit surface area) or catalysts mass (rate of change of moles per unit time per unit catalyst mass).

In all cases, quantification of the intrinsic rate of reaction involves expressing the rate as a function of concentrations (or pressures) of species involved in the reaction, and temperature[2]. Consider a simple reaction involving two species, A and B:

$$A + B \rightarrow R + S$$

The rate of the reaction can be written in terms of any one of the species involved in the reaction. It can be seen from this equation that the species reaction rates are interrelated through the stoichiometry of the reaction. Mathematically, this is as follows:

$$-r_A = -r_B = r_R = r_S \qquad (9.2)$$

The negative sign associated with A and B signifies that these two species are reactants that are consumed in the reaction. The rate for these species is the rate of disappearance. Conversely, for R and S, the products of the reaction, the rate is the rate of generation of the species. Equation 9.2 is stating that the rate of disappearance of A is exactly the same as the rate of generation of R, and so on. It is common to use a *power-law model* to describe the dependence of the reaction rate on concentrations, as shown in equation 9.3:

$$-r_A = kC_A^{\alpha}C_B^{\beta} \qquad (9.3)$$

Here, k is the rate constant, and α and β—the exponents of the concentrations of A and B—are the *orders* of the reaction with respect to A and B, respectively. The overall order of the reaction is the sum of the orders of the reactants, which is $(\alpha + \beta)$ for the previous expression [7]. The intrinsic rate constant and the reaction orders are independent of time and concentration of species. A first-order reaction implies that the rate is directly proportional to the concentration, and a second-order reaction means that the rate is proportional to the square of the concentration. A zero-order reaction does not exhibit any concentration dependence. While these are the most common reaction orders proposed for rate equations, other orders are possible, and it is not necessary for the order to be an integer [7].

The temperature dependence of the reaction rate is incorporated in the rate expression through the rate constant k. This dependence is typically expressed by the Arrhenius[3] expression [6]:

$$k = Ae^{-E_a/RT}$$

2. A catalyst does not participate in the reaction, and catalyst concentration does not appear explicitly in the rate expression. Its influence on the rate is incorporated in the rate constant in the rate expression.

3. Svante Arrhenius, 1903 Nobel Laureate in Chemistry, is one of the first scientists to work on carbon dioxide and greenhouse effect.

In this expression, A is called the *frequency factor*, and E_a is the activation energy for the reaction. Both these parameters are determined experimentally [5].

It should be noted that the units of the rate constant depend on the reaction order. The rate is typically expressed on a volumetric basis, that is, in terms of mole per unit time per unit volume. If the concentrations are expressed in (mol/volume) units, then the units of rate constant are per unit time. On the other hand, if the reaction is zero-order, then the units of rate constant are the same as the units of the rate; that is, mol per unit time per unit volume.

The power-law model offers a useful and convenient expression to describe the rate-dependence on concentration. However, it should be realized that it is not necessarily an accurate and exact description of the changes occurring at the molecular level. Other complex expressions based on postulated reaction mechanisms can be derived and offer quantitative accuracy. However, these expressions will also involve more parameters that need to be determined experimentally with confidence to achieve the desired accuracy.

Determination of the intrinsic kinetics for a reaction involves experimental determination of reaction orders (with respect to species and overall) and the rate constant (including the frequency factor and activation energy). Determination of the intrinsic rate expression constitutes a fundamentally important topic in chemical engineering kinetics. Designing proper laboratory experiments and performing accurate data analysis are critically important for obtaining reliable estimates of rate parameters that, in turn, form the basis of reactor design.

Once the intrinsic kinetics is determined, it is coupled with the reactor behavior to obtain design equations for the reactor. The different reactor types mentioned in Chapter 3, "Making of a Chemical Engineer," are described in more detail and quantitative expressions governing their behaviors are presented next.

9.1.2 Batch and Continuous Reactors

Some chemical reactions are conducted in a batchwise mode [6]: raw material is charged initially into the reactor and the reaction allowed to proceed until such a time that the desired quantity of product is obtained. This is an unsteady state process wherein the conditions within the reactor with respect to the number of moles of various species vary with respect to time. Other conditions, such as the temperature and pressure, may vary depending on the mode of operation. Some reactors are operated *isothermally*, that is, at constant temperature, by either removing the heat from or supplying it to the reactor as needed. Other reactors are operated *adiabatically*, that is, with no heat/energy exchange with the surroundings. The reactor temperature will vary with respect to time in this case. Similarly, particularly for gaseous reactions, pressure may vary with respect to time if the reaction is conducted under constant volume conditions. For gaseous reactions, the pressure variation with the progress of the reaction must to be taken into account when the reaction stoichiometry involves a change in the total number of moles going from the reactants to the products. The component balance shown in equation 9.1 is simplified for a batch reactor by setting the molar flow rate terms to 0, leading to equation 9.4.

$$\frac{dN_i}{dt} = V \cdot r_i \tag{9.4}$$

The reactor may also be operated in a continuous mode, with the raw material fed continuously to the reactor and the product withdrawn continuously. These reactors typically operate under steady-state conditions wherein the reactor conditions remain invariant with respect to time. Equation 9.1 simplifies to equation 9.5 for continuous steady-state reactors.

$$F_{i,in} - F_{i,out} + V \cdot r_i = 0 \tag{9.5}$$

Continuous reactors may also be operated isothermally or adiabatically. Figure 9.2 shows the schematics of a batch reactor and three types of continuous reactors [5]. The batch reactor does not have any influent or effluent streams, whereas all the continuous reactors have both the influent and effluent streams. The stirrer/agitator in the batch reactor indicates that contents are well mixed; that is, there are no spatial variations in conditions with location within the reactor. Similarly, the contents of the first of the continuous reactors—the mixed flow reactor (MFR) or the continuous stirred tank reactor (CSTR)—are well mixed, and uniform conditions exist throughout the reaction vessel. The other two reactors are tubular reactors having the configuration of a long pipe. Obviously, the contents of these types of reactors are not well mixed, and the conditions (concentrations and possibly temperature and pressure) vary as a function of position or location within the reactor. The first of these reactors is essentially an empty pipe or tube through which the reacting fluid flows, typically in what is termed the *plug flow pattern*. This plug flow reactor (PFR) is used for conducting homogeneous gas- or liquid-phase reactions. It is also possible to operate the reactor

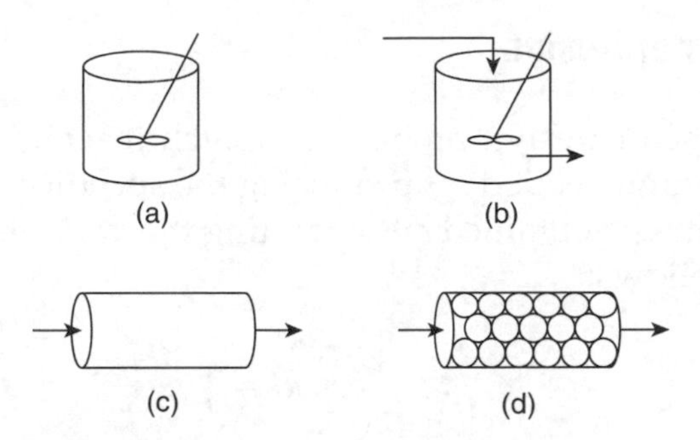

Figure 9.2 Batch and continuous reactors: (a) batch reactor, (b) mixed flow reactor/continuous stirred tank reactor, (c) plug flow reactor, (d) packed bed reactor.
Adapted from: Fogler, H. S., *Elements of Chemical Reaction Engineering*, Fourth Edition, Prentice Hall, Upper Saddle River, New Jersey, 2004.

such that the flow within the reactor is laminar (refer to Chapter 5, "Computations in Fluid Flow"); however, such laminar flow reactors (LFRs) are not as common as the PFRs. As mentioned earlier, many chemical reactions require the use of solid catalysts. The fourth configuration shown in the figure is a packed bed reactor (PBR), a tubular reactor filled with solid catalyst particles operated in plug flow mode for conducting heterogeneous reactions.

It can be seen that the molar flow rates in equation 9.5, or the rate of change of moles in equation 9.4, are determined by the specified production rate. Knowing intrinsic reaction rate r_i allows us to determine the reactor volume or perform other calculations for reactor design based on equation 9.4 for a batch reactor or equation 9.5 for continuous reactors, as described shortly. Note that sometimes reactions are conducted in a semibatch mode wherein one of the reactants is added to another reactant that is already present in the reactor and no product is withdrawn from the reactor until the reaction is completed. Other semibatch modes may include no inflow of reactants but a continuous removal of one or more of the product streams, and many other arrangements involving complex feed/product withdrawal and heating/cooling cycles. Such complex operations require the use of equation 9.1 for design and are not dealt with in this text.

9.1.3 Reactor Design

A typical reactor design problem for batch reactors involves determination of the batch time needed for producing the specified quantity of the product. This batch time is obtained by separating the variables and integrating equation 9.4 as follows:

$$t_{batch} = \int_{0}^{t_{batch}} dt = \int_{N_{i0}}^{N_i} \frac{dN_i}{V \cdot r_i} \tag{9.6}$$

Here, N_{i0} are the initial moles of i. Many of the batch reactors operate under constant volume conditions; that is, the volume of the reactor is fixed. In this case, equation 9.6 can be expressed conveniently in terms of concentrations, since $N_i = V{\cdot}C_i$ for a well-mixed reactor and the rate r_i is a function of concentrations of the species involved in the reaction. For the reaction shown earlier (A + B → R + S), the batch time can be expressed in terms of concentration of A as follows:

$$t_{batch} = -\int_{C_{A0}}^{C_A} \frac{dC_A}{f(C_A, T)} \tag{9.7}$$

The denominator is the rate of disappearance of A ($-r_A$), which is a function of the concentration of A and temperature. The negative sign signifies that A is a reactant that is consumed in the reaction, as discussed in section 9.1.1. C_{A0} and C_A are the concentrations of A at the beginning and the end of the reaction. The specific form of the function depends on the rate parameters describing the intrinsic kinetics of the reaction. For a reaction that is first order with respect to A—that is, $-r_A = f(C_A, T) = k{\cdot}C_A$—integration of equation 9.7 yields the following:

$$t_{batch} = -\frac{1}{k}\ln\frac{C_A}{C_{A0}} \tag{9.8}$$

The result shown in equation 9.8 is based on the assumption that the reaction is conducted isothermally; that is, the temperature remains constant, and hence the rate constant does not vary with time. For adiabatic and other nonisothermal modes of operation, numerical integration of equation 9.7 may be needed for obtaining the batch time.

Equation 9.5 can be used for the design of a CSTR/MFR, as an assumption implicit in equation 9.5 (and the original equation 9.1) is that the contents of the reactor are uniform, which holds true for the CSTR/MFR. The reactor volume can then be determined using equation 9.9 [4]:

$$V_{CSTR} = \frac{F_{A0} - F_A}{-r_A} \quad (9.9)$$

If the reaction does not involve any change in volume, then equation 9.9 can be simplified by writing the molar flow rates as products of the volumetric flow rate and concentration ($F_A = v \cdot C_A$, v being the volumetric flow rate) to obtain equation 9.10:

$$\bar{t}_{CSTR} = \frac{V_{CSTR}}{v} = \frac{C_{A0} - C_A}{f(C_A, T)} \quad (9.10)$$

In this equation, $\bar{t}_{CSTR}$ is the mean residence time of the fluid in the reactor. Depending on the nature of the functional dependence of the rate on concentration, an algebraic or transcendental equation in C_A is obtained as the design equation for the CSTR [4].

Equation 9.5 cannot be used directly for the design of PFR or PBR, as conditions in these reactors show spatial variation. For these reactors, the component material balances are written over a differential element of the reactor, and the reactor volume for PFR is obtained by the integration of the resulting differential equation as shown by equations 9.11 and 9.12, respectively:

$$-\frac{dF_A}{dV} = -r_A \quad (9.11)$$

$$V_{PFR} = -\int_{F_{A0}}^{F_A} \frac{dF_A}{-r_A} \quad (9.12)$$

For a PBR, the rate expression is usually written in terms of catalyst mass, and the equivalent differential and integral forms of the previous two equations are as follows:

$$-\frac{dF_A}{dW} = -r_A' \quad (9.13)$$

$$W_{Catalyst} = -\int_{F_{A0}}^{F_A} \frac{dF_A}{-r_A'} \quad (9.14)$$

Here, r_A' indicates that the rate is expressed in terms of per unit mass rather than per unit volume.

The molar flow rates can be substituted by the products of concentrations and volumetric flow rates, and when the reaction does not involve any volume change, equation 9.12 for the PFR simplifies as follows:

$$\bar{t}_{PFR} = \frac{V_{PFR}}{v} = -\int_{C_{A0}}^{C_A} \frac{dC_A}{f(C_A, \mathrm{T})} \tag{9.15}$$

Here, $\bar{t}_{PFR}$ is the mean residence time in the PFR.

The similarity and equivalence between a batch reactor and a PFR should readily become apparent by a comparison of equations 9.7 and 9.15. The batch time in a batch reactor is equivalent to the mean residence time in the PFR. As with the batch reactor, equation 9.15 may be integrated analytically for isothermal reactors and other situations, while numerical and other techniques may be needed for nonisothermal operations and complex rate equations. A graphical representation of equations 9.7, 9.10, and 9.15 is shown in Figure 9.3.

The batch time or the mean residence time for the PFR is represented by the area under the curve bound by the final and initial concentrations. The mean residence time for the CSTR is the area of the rectangle formed by the sides of the length ($C_{A0} - C_A$) and $-1/r_A$ evaluated at C_A. This area includes the area under the curve as well as the area above the curve bound by the dashed lines. It can be seen that, in this instance, the mean residence time needed for the CSTR operation is greater than the mean residence time needed in the PFR. This translates into a higher volume requirement for the CSTR for the same processing rate, and it may be advantageous to choose a

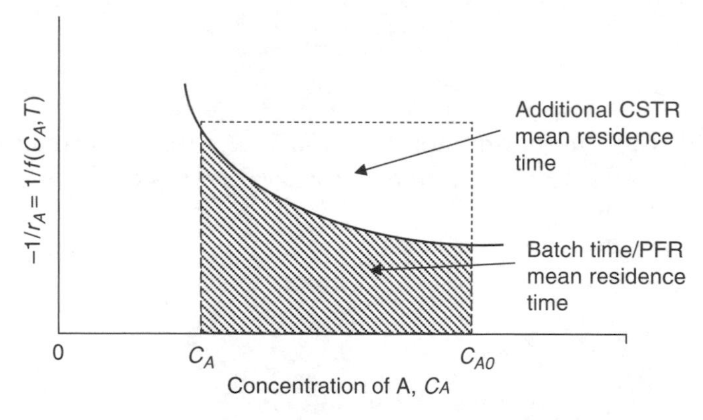

Figure 9.3 Graphical interpretation of equations 9.7, 9.10, and 9.15. CSTR, continuous stirred tank reactor; PFR, plug flow reactor.

PFR to conduct the reaction. Figure 9.3 is a type of *Levenspiel plot* used to obtain the reactor volume, residence time, and batch time for the reactors [5]. A more common representation of the Levenspiel plot involves a graph of $F_{A0}/(-r_A)$ as a function of conversion of A.

9.1.4 Conversion

The discussion in section 9.1.3 assumes that the rate expression is dependent only on the concentration of one reactant, reactant A. However, in general, the rate depends on the concentrations of several other species as well. The component balance in equation 9.1 (and its simplifications discussed earlier) can be written for each species involved in the reaction, yielding as many equations in the species concentrations as the number of species. This will result in a large number of differential equations or algebraic/transcendental equations that need to be solved simultaneously for obtaining the species concentrations and hence determining the reactor volume, residence time, and batch time. The concept of conversion reduces this complexity by expressing concentrations of all the species in terms of a single variable and a single governing equation for reactor design.

Conversion (X) of the *limiting reactant* (reactant that would be consumed completely if the reaction were allowed to run its course) is simply the fraction of the reactant fed to the reactor that undergoes the reaction. If A is the limiting reactant, then in a batch system, we get the following:

$$X_A = \frac{N_{A0} - N_A}{N_{A0}} \tag{9.16}$$

Here, N_{A0} and N_A are the moles of A present initially and those remaining after time t, respectively.

For a continuous reactor, the conversion is described in terms of molar flow rates (F_{A0}, F_A) instead of number of moles (N_{A0}, N_A). For a constant volume system, the conversion is related to the concentrations as follows:

$$X_A = \frac{C_{A0} - C_A}{C_{A0}} \tag{9.17}$$

The concentrations of all the other species (excess reactants, products) can then be expressed in terms of X_A using the reaction stoichiometry The system of equations is now reduced to a single governing equation that can be solved for X_A for a given reactor volume or for V when conversion is specified. For example, if the reaction can be represented by the following equation:

$$\nu_A A + \nu_B B \rightarrow \nu_R R + \nu_S S \quad (9.18)$$

then, the concentration of B can be expressed in terms of conversion of the limiting reactant A by the following equation:

$$C_B = C_{A0}\left(\psi_B - \frac{\nu_B}{\nu_A} X_A\right) \quad (9.19)$$

Here, ψ_B is the ratio of the initial concentrations of B and A ($\psi_B = C_{B0}/C_{A0}$). In general, the concentration of any species i is expressed using a similar expression [8]:

$$C_i = C_{A0}\left(\psi_i - \frac{\nu_i}{\nu_A} X_A\right) \quad (9.20)$$

Here, ν_i represents the stoichiometric coefficient of the species i in the reaction, and ψ_i is the ratio of the initial concentrations of i and A. It should be noted that the validity of equation 9.20 is restricted to constant-volume systems, and the reaction equation is written in the form *Products* – *Reactants* = 0; thus equation 9.18 is rearranged to read:

$$\nu_R R + \nu_S S - (\nu_A A + \nu_B B) = 0 \quad (9.21)$$

Typically, the approach to kinetic analysis involves developing a *stoichiometric table* [4], wherein the concentrations of various species are expressed in terms of concentration of the limiting reactant, its conversion, and initial concentration ratios (at time $t = 0$ for the batch reactor, and inlet of the reactor for PFR/CSTR). This development leads to equations 9.22, 9.23, and 9.24 that are the design equations for the batch reactor, CSTR, and PFR, respectively [4].

$$t_{batch} = N_{A0}\int_0^{X_A} \frac{dX_A}{f(X_A,T)\cdot V} \quad (9.22)$$

$$V_{CSTR} = \frac{F_{A0}\cdot X_A}{f(X_A,T)} \quad (9.23)$$

$$V_{PFR} = F_{A0}\int_0^{X_A} \frac{dX_A}{f(X_A,T)} \quad (9.24)$$

Depending on the information provided and the nature of the problems, these equations can be used to obtain reactor volumes, batch times, conversions in reactors of specified volumes, processing rates, and so on.

9.1.5 Other Considerations

The reactor analysis and design acquires another level of complexity when the reactor is operated under nonisothermal conditions; that is, the temperature of the reactor varies with time or position in the reactor, mostly due to the substantial heat effects associated with most reactions. As the temperature varies, the rate constant also changes according to the Arrhenius equation, and a single governing equation is not sufficient to solve for two variables (T and X_A or T and V). The second equation needed is obtained by performing an energy balance on the system.

The simple conversion-concentration relationship described by equation 9.20 is valid only for a constant volume system. Gas-phase reactions involving changes in pressure, temperature, or the number of moles experience a volume change, and the conversion is not merely a function of concentration but also of these other variables. Additional governing equations are required to describe these situations.

It can be seen that a chemical engineer will be required to solve problems in chemical engineering kinetics that range from simple linear equations to highly complex, multiple, simultaneous differential equations. The representative examples presented in section 9.2 utilize the mathematical relationships that are obtained when the concepts of chemical engineering kinetics are applied to different situations. The development of these mathematical relationships itself requires an in-depth knowledge of concepts and is not covered in this text.

9.2 Basic Computational Problems

Determination of intrinsic kinetics forms an essential and important component of chemical engineering kinetics. Typically, such determinations are carried out by bench-scale batch experiments wherein the concentrations of species are monitored as a function of time and the concentration-time data are subjected to analysis. Example 9.2.1 illustrates such analysis conducted for the determination of rate expression.

Example 9.2.1 Determination of Rate Constant

A laboratory experiment is conducted to determine the rate constants for a reaction 2A → R. It is known that the reaction is second order with respect to A. The experiment involved measuring the concentration of the reactant A as a function of time in a batch vessel. The following is the data obtained. What is the value of the rate constant for the reaction?

Time, h	0	0.02	0.04	0.06	0.1	0.15	0.2	0.225	0.275
C_A, mol/L	1.00	0.83	0.71	0.62	0.50	0.40	0.33	0.31	0.27

Solution

The concentration-time relationship for a second-order reaction follows [3]:

$$\frac{1}{C_A} - \frac{1}{C_{A0}} = k_2 t \tag{E9.1}$$

Here, k_2 is the rate constant for a second-order reaction having the units of (concentration · time)$^{-1}$, L/mol h in this case.

As indicated by equation E9.1, the inverse of the concentration of A varies linearly with time. A graph of $1/C_A$ versus t will have the rate constant k_2 as its slope and $1/C_{A0}$ as the intercept. The rate constant can thus be obtained by a linear regression between $1/C_A$ and t.

Solution (using Excel)

The solution steps using Excel are as follows:

1. Enter the raw data in the spreadsheet: time values in cells A3 through A11 and the concentration values in cells B3 through B11.
2. Calculate the inverse of concentration in cell B3 into cell C3 by entering the formula =1/B3. Copy the formula into cells C4 through C11 to calculate $1/C_A$.
3. Select the data in cells A3 through A11 and C3 through C11 by highlighting these cells: cells A3 through A11 are selected using the mouse, and then C3 through C11 are selected while pressing the Ctrl key to skip column B.
4. Create a graph by selecting Insert from the command menu bar, and then the X-Y scatter plot from the dropdown menu or by using the appropriate buttons. This plots data in cells C3 through C11 ($1/C_A$) as a function of data in cells A3 through A11 (t).
5. Add various chart elements (axes titles, chart title) using the Design tab.

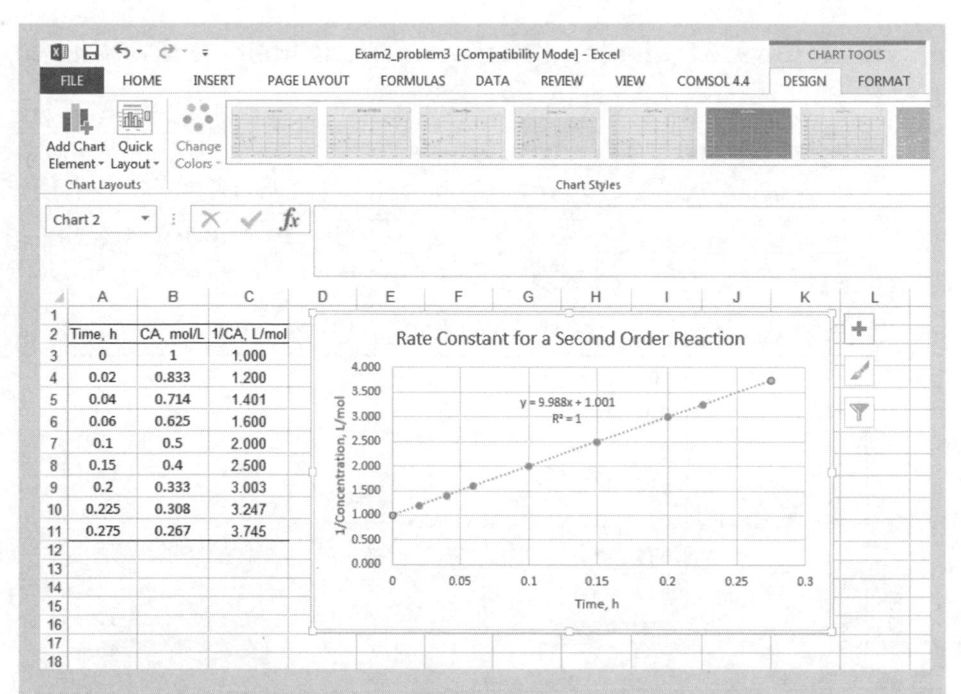

Time, h	CA, mol/L	1/CA, L/mol
0	1	1.000
0.02	0.833	1.200
0.04	0.714	1.401
0.06	0.625	1.600
0.1	0.5	2.000
0.15	0.4	2.500
0.2	0.333	3.003
0.225	0.308	3.247
0.275	0.267	3.745

Figure 9.4 Excel solution to example 9.2.1.

6. Insert a linear regression line by right-clicking on the data points, which opens a dialog box. Clicking the appropriate button inserts the regression line equation and correlation coefficient on the chart, as shown in Figure 9.4.

Solution (using Mathcad)

The Mathcad solution is shown in Figure 9.5.

The solution steps are as follows:

1. Enter the data in a matrix named RateData by first typing RateData: and then clicking on the matrix button. This opens a dialog box in which the number of rows (9) and columns (2) are entered, and the dialog box is closed by clicking OK. Enter the data with the first column having time values and the second column the concentration values.

2. Calculate the inverse of concentration values entered in the second column of RateData by typing InvConc:1/RateData<Ctrl+6> and then entering 1 in the placeholder. (The key combination <Ctrl+6> creates an index for the columns in variable RateData. The indices start from 0 for the first column, 1 for the second column, and so on.) The inverse values can be seen by typing InvConc=.

(Continues)

Example 9.2.1 Determination of Rate Constant (*Continued*)

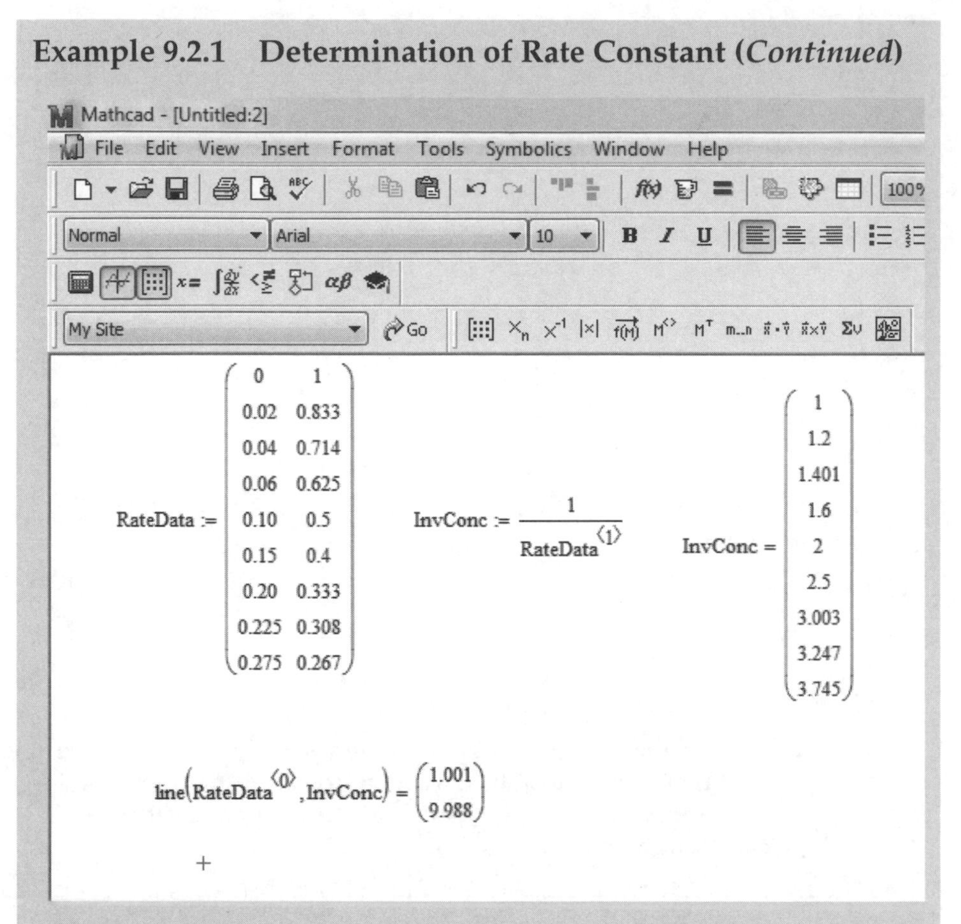

Figure 9.5 Mathcad solution to example 9.2.1.

3. Use the line function to perform linear regression between the time (first column of RateData) and inverse concentration (InvConc). The line function has two arguments: the first one is the independent variable (the first column of RateData in this case), and the second is the dependent variable (InvConc). The keystrokes in this case are line(RateData<Ctrl+6>[space],InvConc), and then enter 0 in the index of RateData. Typing = after the closing parenthesis yields the solution consisting of a vector with two elements. The top number represents the intercept, and the bottom number represents the slope.

As can be seen from the figure, the rate constant value is 9.988 L/mol h, the same value obtained using Excel.

The material balance equation is sufficient for the design and analysis of a CSTR/MFR when heat effects need not be taken into account. However, an energy balance equation is also required when the heat effects are significant and need to be considered. For a given reactor volume, the material balance and the energy balance yield two different equations containing the conversion and the temperature, and these equations need to be solved simultaneously. Example 9.2.2 deals with this situation and existence of multiple solutions for an adiabatic operation of a CSTR.

Example 9.2.2 Conversion in a CSTR

The material balance and energy balance for a CSTR for a first-order reaction are given by equations E9.2 and E9.3, respectively [3]. What is the conversion in the reactor? What is its operating temperature?

$$X_{MB} = \frac{Ae^{-E_a/RT} \cdot \bar{t}}{1 + Ae^{-E_a/RT} \cdot \bar{t}} \tag{E9.2}$$

$$X_{EB} = \frac{C_{PA}\left(T - T_0\right)}{-\Delta H_{rxn}} \tag{E9.3}$$

Here, C_{PA} is the mean specific heat capacity of the reactor contents. $\bar{t}$ is the mean residence time of the material in the reactor—that is, the time it spends in the reactor—obtained by dividing the reactor volume by the volumetric flow rate (V/v). T_0 is the inlet temperature (temperature of the feed stream). X_{EB} and X_{MB} are the conversions using the energy balance and the material balance on the reactor, respectively. ΔH_{rxn} is the heat of reaction.

The data for the reactor are as follows:

$T_0 = 293$ K, $C_{PA} = 810$ J/mol K, $\Delta H_{rxn} = -200$ kJ/mol, $V = 10$ L, $v = 0.2$ L/s

$A = 1.8 \times 10^5\ \text{s}^{-1}$, $E_a = 50$ kJ/mol

(*Continues*)

Example 9.2.2 Conversion in a CSTR (*Continued*)

Solution (using Mathcad)

The first step in the solution using Mathcad is plotting both X_{EB} and X_{MB} as functions of the temperature T. The values of the variables are specified first, followed by specifications of the range of the variable T, and the definitions of the two conversions. Both these functions are plotted against T using the Mathcad Graph utility. Figure 9.6 shows the plots generated.

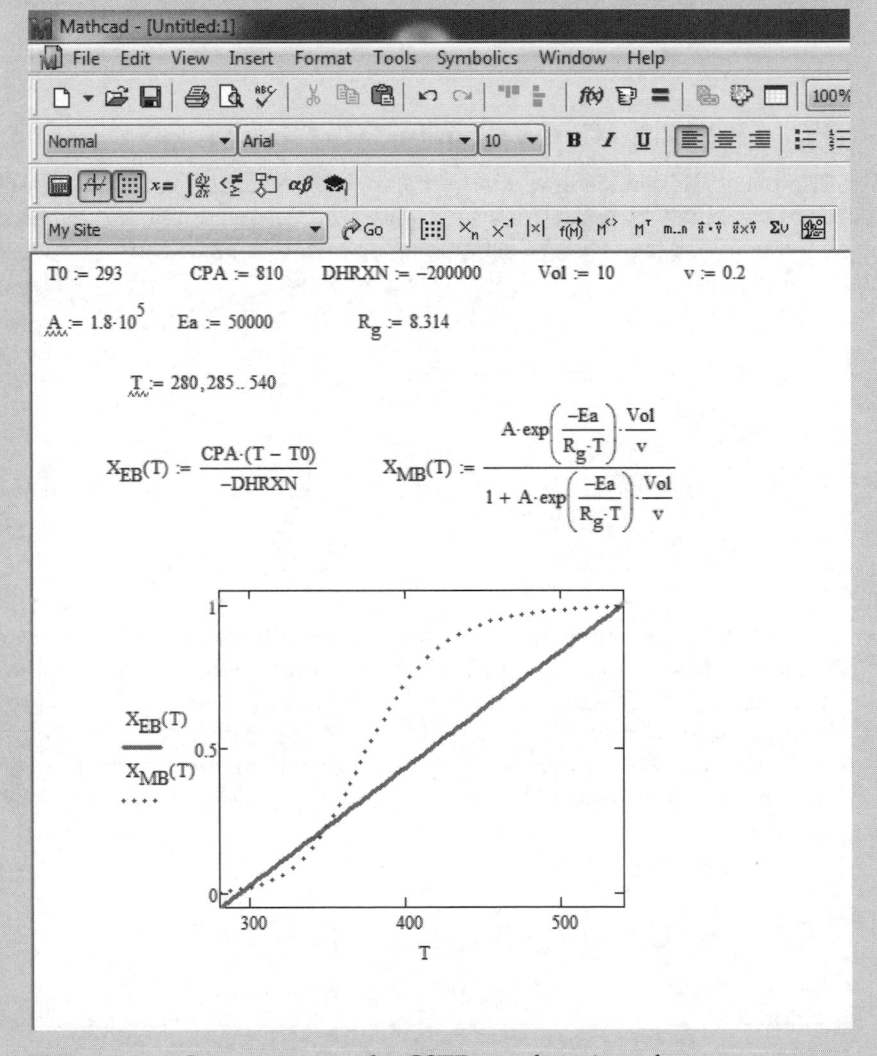

Figure 9.6 Conversions in the CSTR as a function of temperature.

As can be seen from the figure, the two curves intersect in three locations, implying that there are three possible solutions that satisfy the two equations. Examination of the three points reveals a very low, an intermediate, and a very high conversion. Since the objective is typically to convert as much of the reactant to the product as possible, the very high conversion is the desired operating point. The exact solution for conversion and temperature is obtained using the solve block in Mathcad, as shown in Figure 9.7.

The procedure consists of providing an initial guess for temperature (540), specifying the objective function (equality of conversions), and then letting the program find the value of T that satisfies the objective function. As can be seen from the figure, the conversion will be 0.992 at the reactor temperature of 538 K, meaning that 99.2% of the reactant fed to the reactor will be converted to the desired product.

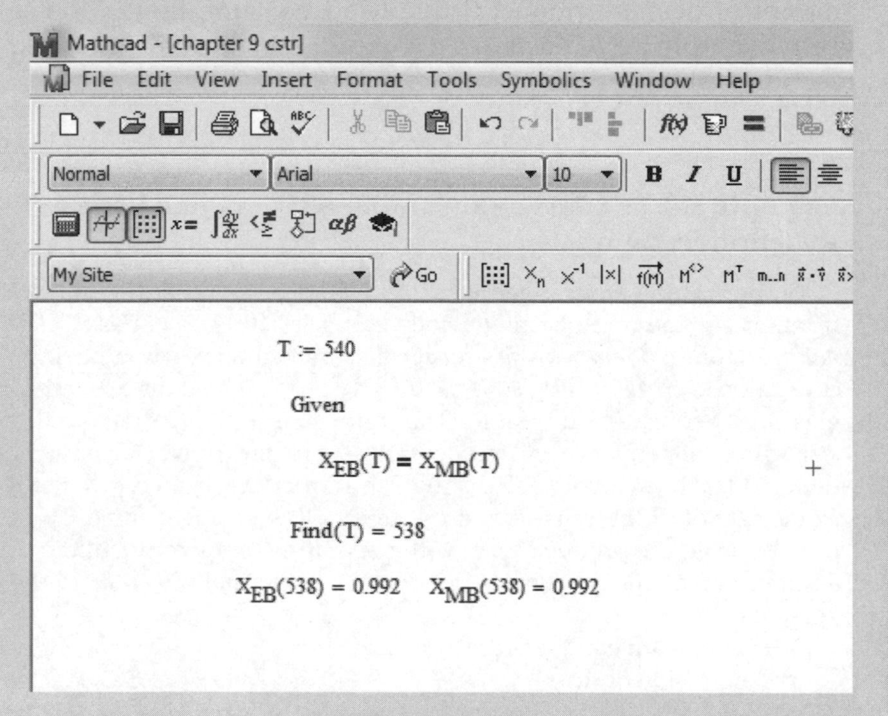

Figure 9.7 Exact solution at the high conversion point.

(Continues)

Figure 9.6, with its multiple intersection points for the energy balance and material balance conversion-time functions, indicates that *multiple steady-state* points are feasible while operating a CSTR. How the conditions are manipulated to operate at a particular point forms an interesting discussion in the course on chemical engineering kinetics.

Most of the reacting systems encountered in practice exhibit a much higher level of complexity than a simple first-order reaction. The desired reaction is invariably accompanied by an undesirable side reaction, which has significant impact on the economics of the process and increases the complexity of separations postreaction. These undesirable reactions include a parallel reaction that the reactant undergoes ($A \rightarrow S$, when $A \rightarrow R$ is desired) or a series reaction ($A \rightarrow R \rightarrow S$, which reduces the amount of desired product R due to its consumption in the second reaction). Example 9.2.3 illustrates a computational problem related to a series reaction system.

Example 9.2.3 Conversion and Selectivity in a Series Reaction System

Isopropenyl allyl ether isomerizes into allyl acetone according a first-order reaction with a rate constant given by $k_1 = 5.4 \times 10^{11} \exp(-123000/R_gT)\ s^{-1}$. Allyl acetone undergoes a first-order decomposition reaction having a rate constant of $k_2 = 4.9 \times 10^{11} \exp(-131000/ R_gT)\ s^{-1}$, where the activation energies are expressed in J/mol. The rate constant equations reflect the Arrhenius dependence on temperature, with the universal gas constant denoted by the symbol R_g to distinguish it from the desired product R. The isomerization reaction is carried out in a PFR, with the desired $C_{R,max}/C_{A0}$ ratio of 0.8 in the product. What is the operating temperature of the reactor, assuming that the reaction is conducted isothermally? What is the conversion of A?

The relevant equations are as follows [8]:

Conversion of A:

$$X_A = 1 - \exp(-Da) \tag{E9.4}$$

Concentration of R at the reactor outlet:

$$\frac{C_R}{C_{A0}} = \frac{k_1/k_2}{1-k_1/k_2}\left[\exp(-Da) - \exp\left(-\frac{Da}{k_1/k_2}\right)\right] \tag{E9.5}$$

In these equations, Da is a dimensionless number known as the *Damköhler number*,[4] which is the product of the rate constant and mean residence time in a continuous reactor for a first-order reaction. It is also known that the ratio of maximum concentration of R, $C_{R,max}$ to the inlet concentration of A is related to the ratio of the rate constants by equation E9.6.

$$\frac{C_{R,\max}}{C_{A0}} = \left(\frac{k_1}{k_2}\right)^{\left(\frac{1}{1-k_1/k_2}\right)} \tag{E9.6}$$

Solution algorithm

The left side of equation E9.6 is specified to be 0.8. This specification is used to obtain the ratio of the rate constants k_1/k_2. The needed temperature can be calculated from equation E9.7.

$$\frac{k_1}{k_2} = \frac{5.4 \cdot 10^{11} \exp(-123000 / R_g T)}{4.9 \cdot 10^{11} \exp(-131000 / R_g T)} \tag{E9.7}$$

Equation E9.5 is used to calculate the Damköhler number, Da, which then is used to obtain the conversion of A from equation E9.4.

Solution (using Mathcad)

The solution using Mathcad is shown in Figure 9.8.

As can be seen from the figure, the solution involves using a number of solve blocks. The ratio of rate constants is obtained from solving equation E9.6 to be 12.216. This leads to the reaction temperature of 400 K from equation E9.7. The Damköhler number is computed to be 2.726 from equation E9.5, and finally the conversion of A is 0.935 from equation E9.4.

(*Continues*)

---

4. The Damköhler number signifies the ratio of the reaction rate of a species in a reactor to its flow rate through the reactor, both based on inlet concentration of the species.

Example 9.2.3 Conversion and Selectivity in a Series Reaction System (*Continued*)

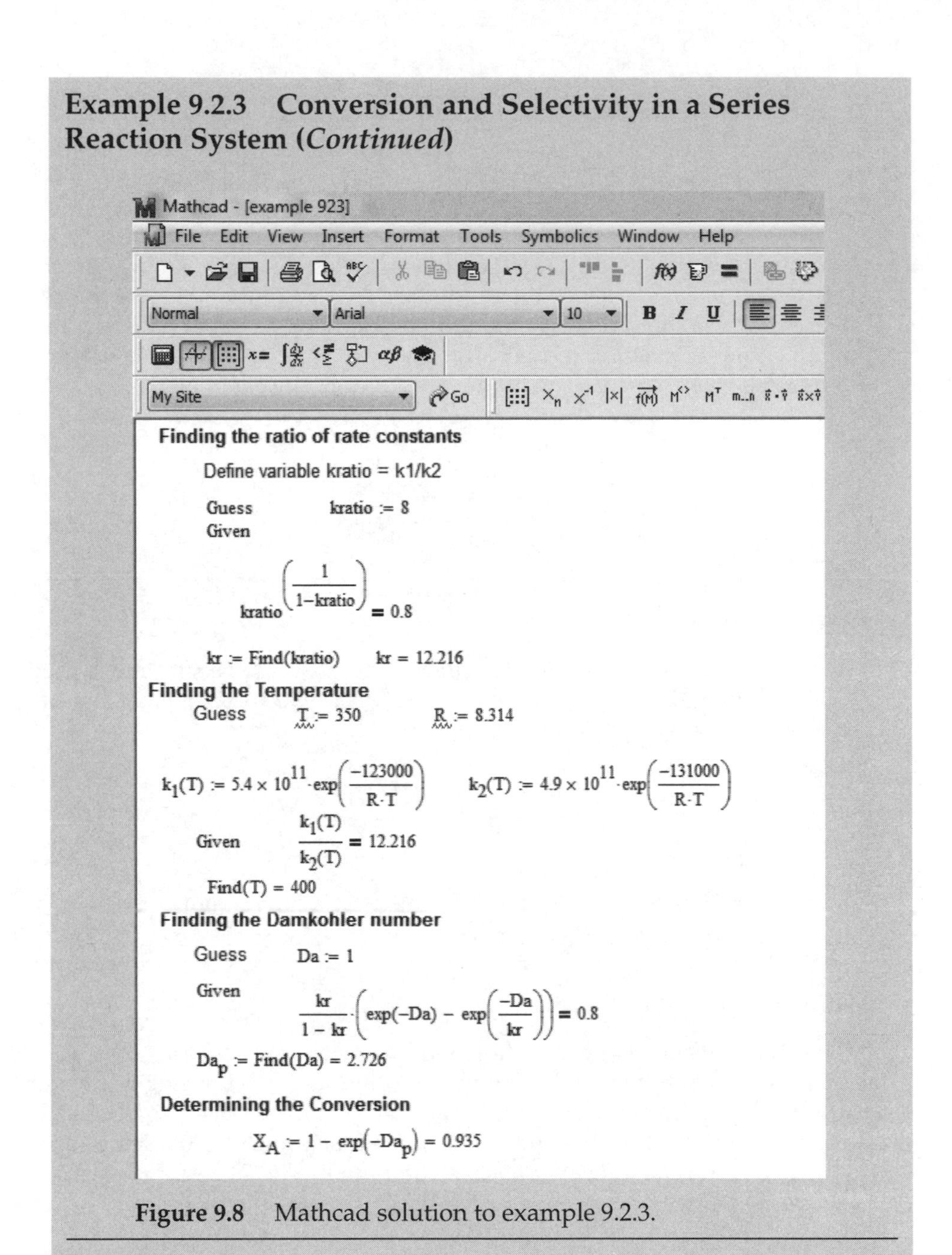

Figure 9.8 Mathcad solution to example 9.2.3.

Solution (using Excel)

The solution using Excel involves use of multiple Goal Seek functions. The resulting solution is shown in Figure 9.9.

It can be seen that both Mathcad and Excel give similar solutions. Excel yields a value of 401 K for the temperature, and the conversion is 0.939. Both these values are very close to the Mathcad solution of 400 K for temperature and 0.935 for conversion.

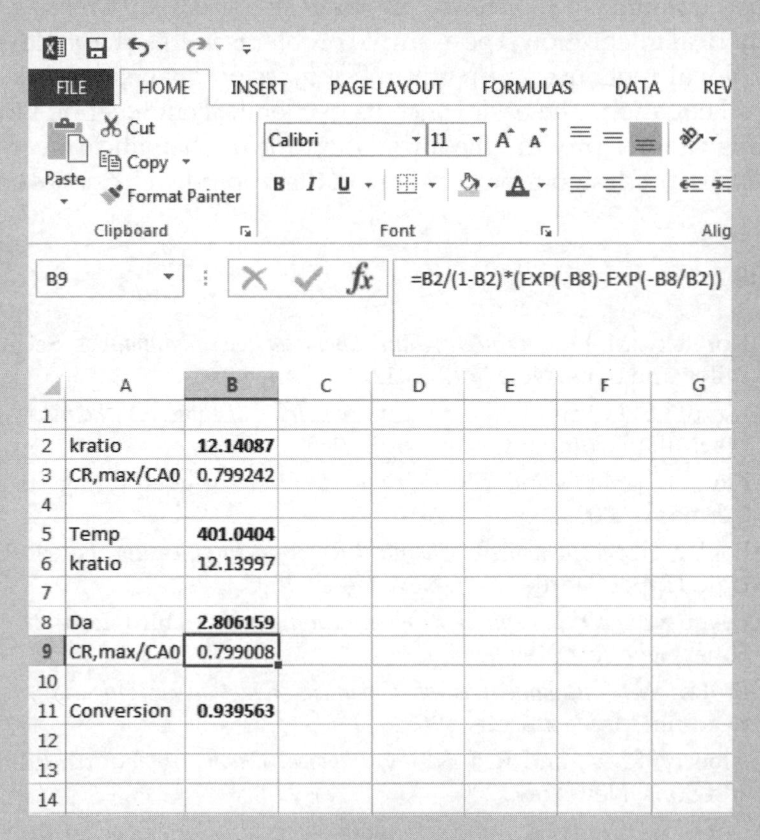

Figure 9.9 Excel solution to example 9.2.3.

These three examples offer a glimpse into the nature of computations that a chemical engineer will typically conduct in dealing with topics related to chemical engineering kinetics. Additional practice problems at the end of the chapter provide students with further perspective on these problems.

9.3 Summary

Chemical engineering kinetics combines the information about the rate of reaction with reactor characteristics to develop optimal reactor design that can make the process economical. A wide variety of computational problems are encountered by a student and a practitioner of the field of chemical engineering kinetics. These problems range from simple linear equations to transcendental and ordinary as well as partial differential equations and numerical integration. The example problems in this chapter describe the computational tools used to perform a regression analysis and solve a transcendental equation. The basic concepts in chemical engineering kinetics described in this chapter provide the foundation for mathematical treatment of reactor analysis and design topics covered in the upper-level kinetics courses.

References

1. Koretsky, M. D., *Engineering and Chemical Thermodynamics,* Second Edition, John Wiley and Sons, New York, 2012.
2. Maron, S. H., and C. F. Prutton, *Principles of Physical Chemistry*, Fourth Edition, MacMillan Company, New York, 1965.
3. Fink, J. K., *Physical Chemistry in Depth,* Springer-Verlag, Berlin Heidelberg, Germany, 2009.
4. Fogler, H. S., *Elements of Chemical Reaction Engineering*, Fourth Edition, Prentice Hall, Upper Saddle River, New Jersey, 2005.
5. Levenspiel, O., *Chemical Reaction Engineering,* Third Edition, John Wiley and Sons, New York, 1999.
6. Brötz, W., *Fundamentals of Chemical Reaction Engineering,* Addison-Wesley, Reading, Massachusetts, 1965.
7. Alberty, R. A., and R. J. Silbey, *Physical Chemistry*, Fourth Edition, John Wiley and Sons, New York, 2004.
8. Doraiswamy, L. K., and D. Üner, *Chemical Reaction Engineering: Beyond the Fundamentals*, CRC Press, Boca Raton, Florida, 2013.

Problems

9.1 The conversion (x_A)-time (t) relationship for a certain reaction is described as follows:

$$t/\tau = 1 - 3\,(1 - x_A)^{2/3} + 2\,(1 - x_A)$$

Here, τ is the time needed for complete conversion ($x_A = 1$). What is the conversion at (i) $t = 0.25\ \tau$, (ii) $t = 0.5\ \tau$?

9.2 The rate of the gas phase reaction $2NO_2 \rightarrow N_2O_4$ can be monitored by monitoring the total pressure of the system. The total pressure-time data for an experimental run follows:

t, s	0	10	20	30	40	50
P_t, atm	3.05	2.23	2.03	1.91	1.80	1.74

Calculate the rate constant if the total pressure-time relationship is described as follows:

$$\frac{1}{2P_t - P_o} = \frac{1}{P_o} + 2k_p t$$

Here, P_t is the total pressure at time t, P_0 is the initial pressure and k_P is the rate constant in $atm^{-1}s^{-1}$.

9.3 The conversion-time relationship for a reaction is described as follows:

$$X_A = \theta(1 - e^{-(1/\theta)})$$

Here, θ is the dimensionless time. At what θ will the conversion be 0.75? When will you get complete conversion?

9.4 Reversible reactions, such as those that can be represented by the equation of the type $A + B \leftrightarrows 2R$ are equilibrium limited; that is, the conversion is limited by the equilibrium between the forward and reverse reactions. The equilibrium conversion X_{Ae} for such reactions can be calculated from the equilibrium constant K_{eq}, which is a function of temperature. At a particular operating condition, the relationship between the equilibrium constant and equilibrium conversion for the above reaction was found to be described as follows:

$$K_{eq} = \frac{4X_{Ae}^2}{(1 - X_{Ae})(1.5 - X_{Ae})}$$

What is the equilibrium conversion, if $K_{eq} = 64$?

9.5 The radioactive decay of an isotope is a first-order process; that is, the decay rate at any instant (measured as disintegrations per time or count rate) depends on the number of atoms of the radioactive isotope present at that instant. The number of atoms of the isotope present at any time t (N) are related to the number of atoms present at time $t = 0$ (N_0) by the equation $N = N_0 e^{-\lambda t}$, where λ is the characteristic constant, which is obtained from the count rate-time data by a linear regression between ln(*Count Rate*) and time. The following data were obtained for a radioactive isotope:

t, hr	Count Rate, min^{-1}
0	550.33
1	549.83
5	549.00
10	548.17
20	545.83
30	544.17

What is the characteristic constant λ for this isotope? Identify the isotope from the following probable choices based on the half-life obtained from the preceding data. The half-life of the isotope is related to the characteristic constant λ by $t_{1/2} = \ln(2)/\lambda$. Half-life data is ^{89}Sr – 53 days, ^{95}Zr – 65 days, ^{90}Yr – 61 hr, and ^{95}Nb – 35 days.

9.6 For catalytic reactions, the concentration profile of a reactant in a spherical particle (concentration as a function of radial position) is often described by the following equation:

$$C_A = C_{AS}\frac{R}{r}\frac{sinh\left(\phi\frac{r}{R}\right)}{sinh(\phi)}$$

Here, C_{As} and C_A are the concentrations at the surface and radial position r, respectively. R is the radius of the spherical particle, and ϕ is the characteristic parameter that is called the *Thiele modulus*. Laboratory experimentation revealed that the concentration decreased by 60% from the surface to the middle of the particle. What is the value of the Thiele modulus?

9.7 Calculate the activation energy for a reaction if the rate constant doubles when the temperature is raised from 300 K to 350 K. What further rise in temperature is needed for doubling the rate constant over its value at 350 K?

9.8 The energy balance for an adiabatic MFR yields the following equation:

$$0.02(T - T_0) = \frac{0.06exp\left(15000\left(\frac{1}{300} - \frac{1}{T}\right)\right)}{0.06exp\left(15000\left(\frac{1}{300} - \frac{1}{T}\right)\right) + 1}$$

Here, T is the reactor temperature and T_0 the temperature of the inlet stream. What is (are) the reactor temperature(s) when the inlet temperature is 290 K? 295 K? If each side of the equation is also equal to the conversion in the reactor, what is (are) the conversion(s)? What operating temperatures are preferable? (Hint: First create a plot of both sides of the equation as a function of temperature).

9.9 The following data were obtained for conversion of A in a reversible reaction $A + B \rightleftarrows R + S$:

Time, min	Conversion
0	0
10	0.05
20	0.09
30	0.13
40	0.167
50	0.2
60	0.23
70	0.26
80	0.29
90	0.31
∞	0.80

Determine the rate constant k if the conversion-time relationship can be expressed by the following equation:

$$ln\frac{X_{Ae}-(2X_{Ae}-1)X_A}{X_{Ae}-X_A}=2k\left(\frac{1}{X_{Ae}}-1\right)C_{A0}t$$

C_{A0}, the initial concentration of A is 1 mol/L.

9.10 The concept of a *residence time distribution* is critically important for describing the nonideality in the flow behavior of continuous reactors. A dimensionless number called *Peclet number* (*Pe*) is often used for the mathematical description of the nonideality. The mean residence time, $\bar{t}$, and the variance, σ^2, in the residence time distribution are dependent on the Peclet number as follows:

$$\sigma^2=\bar{t}^2\left(\frac{2}{Pe}-\frac{2}{Pe^2}(1-e^{-Pe})\right)$$

The mean residence time and the standard deviation (σ) for a vessel were found to be $5{\cdot}10^5$ s and 305 s, respectively. What is the Peclet number? What would be the error in *Pe* if the second term in the bracket is neglected?

EPILOGUE

Students in an engineering program (or for that matter, any professional program) have typically only the vaguest notions of what they will be doing after graduation from the program, or even what they will be learning in the program until they are well into it.[1] It is the author's expectation and belief that the students persevering through the chapters of the book to reach this point will have a much clearer insight into and appreciation of their chosen field. They will be better informed and have a more concrete conceptual understanding of chemical engineering from practically the beginning of the program.

As the students progress through the subsequent courses of their chemical engineering program, they will delve deeper into the concepts introduced in this book. They will also encounter computational problems of increasingly greater complexity; and, they will also discover that they are wielding exactly the same tools as those learned in this book to solve the problems! They will confidently exploit the powerful capabilities of the software programs and process simulation software[2] to solve challenging problems effectively, efficiently, and quickly.

Chemical engineering is a rigorous discipline that places much demand on its student and, in return, provides unparalleled rewards to one who is willing to put in the required efforts. The author is confident that this book will continue to serve the readers as they transition from freshmen to seniors to graduates and beyond in their careers.

1. Based on the author's experience with the students, as well as his own experience.
2. Introduced in the appendixes.

APPENDIX A

Introduction to Mathematical Software Packages

Chapter 4, "Introduction to Computations in Chemical Engineering," introduced a variety of mathematical problems encountered by a chemical engineer in his/her academic and professional career. Chapters 5 through 9 presented some of these problems and computations executed using Excel and Mathcad. Excel is, of course, ubiquitous on personal computers as a part of the Microsoft Office Suite. The other software, Mathcad, is not as universally available. Nevertheless, several other software packages that can offer comparable capabilities are invariably available, both in academic institutions and organizations beyond. All these packages offer exceptional PC-based computational power to solve practically any problem in the chemical engineering field. These software packages have essentially eliminated any need for an individual to develop customized programs in a higher-level computer language (such as Fortran) by developing specific tools to address particular computational needs [1]. A brief introduction to other software packages (apart from Mathcad) is provided in this appendix. These packages include POLYMATH (www.polymath-software.com), MATLAB (www.mathworks.com/products/matlab), Maple (www.Maplesoft.com), and Mathematica (www.Wolfram.com). This appendix includes a solution of one of the example problems discussed in Chapter 9, "Computations in Chemical Engineering Kinetics," using two of these software packages, followed by a brief comparative discussion of some of the salient features of the packages.

A.1 Example 9.2.2 Revisited

Example 9.2.2 involved determining the conversions in an adiabatic continuous stirred tank reactor (CSTR) by solving the mass and energy balance equations simultaneously. The problem is restated here:

> The material balance and energy balance for a CSTR for a first-order reaction are given by equations A.1 and A.2, respectively. What is the conversion in the reactor? What is its operating temperature?

$$X_{MB} = \frac{Ae^{-E_a/RT} \cdot \bar{t}}{1 + Ae^{-E_a/RT} \cdot \bar{t}} \tag{A.1}$$

$$X_{EB} = \frac{C_{PA}\left(T - T_0\right)}{-\Delta H_{rxn}} \tag{A.2}$$

Here, C_{PA} is the mean specific heat capacity of the reactor contents. $\bar{t}$ is the mean residence time of the material in the reactor—that is, the time it spends in the reactor—obtained by dividing the reactor volume by the volumetric flow rate (V/v). T_0 is the inlet temperature (temperature of the feed stream). X_{EB} and X_{MB} are the conversions using the energy balance and the material balance on the reactor, respectively. ΔH_{rxn} is the heat of reaction.

The data for the reactor are as follows:

$T_0 = 293$ K, $C_{PA} = 810$ J/mol K, $\Delta H_{rxn} = -200$ kJ/mol, $V = 10$ L,

$v = 0.2$ L/s $A = 1.8 \times 10^5$ s^{-1}, $E_a = 50$ kJ/mol

The solution technique for the problem using Mathcad was illustrated in Chapter 9. Readers should by now be familiar with the technique to solve the problem using Excel, so that solution is not discussed here. Solutions using POLYMATH and MATLAB are presented in the following sections. The installation of these programs on PCs is a straightforward matter and hence is not discussed here. The program is launched as any other program installed on a PC, by clicking on the icon or selecting from the list of programs.

A.1.1 POLYMATH Solution

Launching the POLYMATH program brings up the screen shown in Figure A.1.

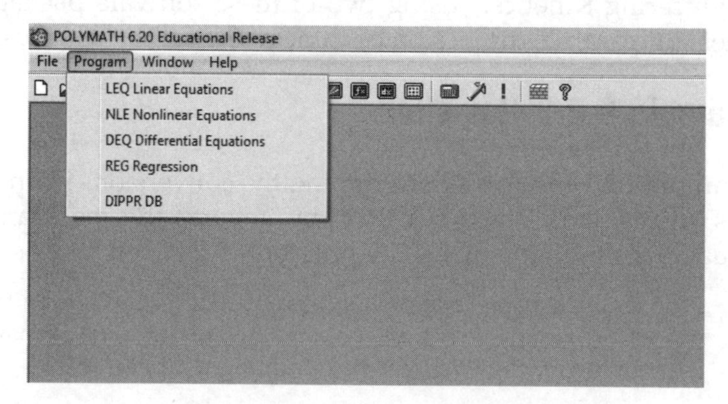

Figure A.1 Opening screen of POLYMATH.

As can be seen from the figure, POLYMATH can be used to solve linear equations, nonlinear equations, and ordinary differential equations as well as perform regression analysis—the options are available under the Program command. The last option seen, DIPPR DB, refers to the database of physical properties through the Design Institute for Physical Properties (DIPPR) of the American Institute of Chemical Engineers (AIChE). The example problem involves a nonlinear equation, and hence that option is chosen from the Program command dropdown menu.

Clicking the NLE Nonlinear Equations option brings up another screen in POLYMATH where users can enter the needed equations. The nonlinear equations are entered by clicking on the appropriate icon, which brings up a dialog box where the equation can be entered as a function of the variable to be solved for. The problem specification typically requires a number of other auxiliary linear equations to be entered (such as those providing the data for the problem), and these can be entered by clicking the icon for auxiliary equation or by simply typing the equation.

Once all the equations are entered, lower and upper limits are specified for the variable, and the program is run by clicking the Run icon or by selecting it from the menu. The user can select an option to draw a graph of the solution. This procedure was followed for the example problem, and the resultant solution is shown in Figure A.2.

The salient features of the procedure and solution are as follows:

1. The data are entered as auxiliary equations by simply typing a variable name and its value.
2. The nonlinear equation is entered as a function of temperature. The function is entered as $f(T) = X_{MB} - X_{EB}$. (Here, X_{MB} and X_{EB} represent the right hand sides of equations A.1 and A.2, respectively, as seen from the two equations and Fig. A.2).
3. A search space is defined by specifying the minimum and maximum temperatures.
4. POLYMATH performs the computations and plots the function as shown in the plot. It can be seen that the function value is 0 at three temperatures.
5. In addition to the graph, POLYMATH outputs the three temperature values. Two of the three solutions are visible in the figure. The third solution (solution #3 of 3) is ~538 K, the same as that obtained using Mathcad. The conversion is 0.99 at this temperature.

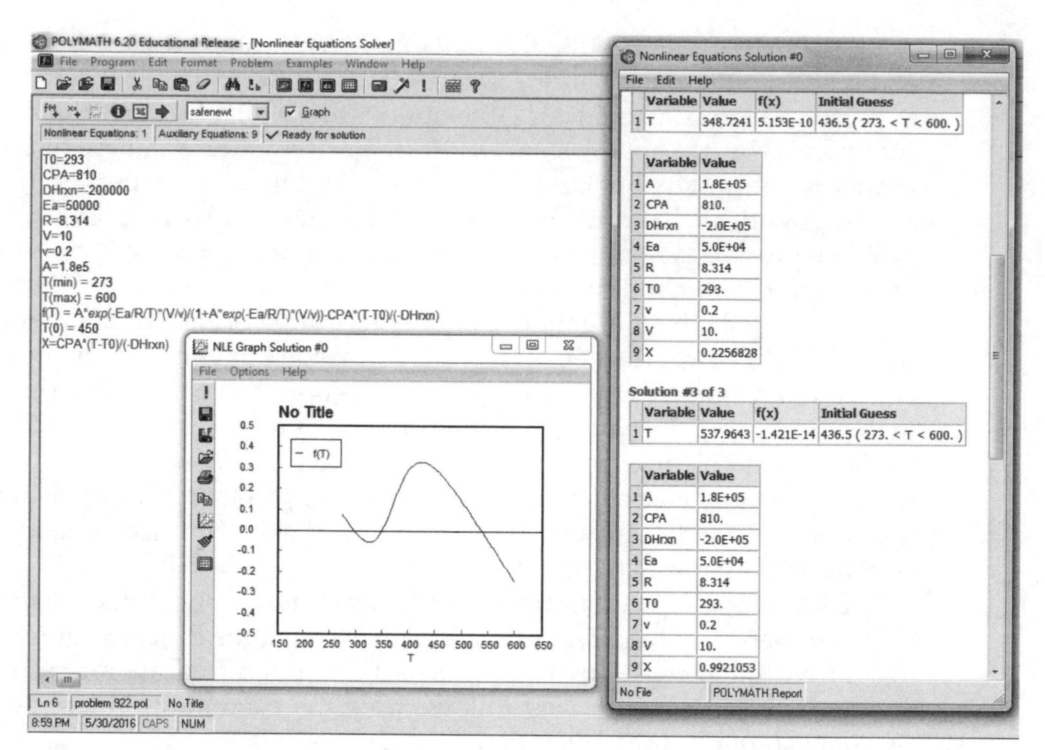

Figure A.2 POLYMATH solution to example 9.2.2.

A.1.2 MATLAB Solution

Launching the MATLAB program brings up the screen shown in Figure A.3. The opening screen of MATLAB is considerably more detailed than those of the other programs encountered until now. The menu bar has a number of icons and options, and the workspace is divided into four areas. The largest area is titled Command Window, and this is the main workspace for entering data and instructions. The space to the left deals with the organization of the files and folders, and the two spaces to the right contain the information about the current problem being worked on. This information involves a command history at the bottom and the tracking of variables above it.

The problem is solved by entering the values of the variables—the specified data for the problem. Once all the values are entered, the solution is obtained by typing the following line:

```
T=solve(A*exp(-Ea/R/T)*(V/v)/(1+A*exp(-Ea/R/T)*(V/v))==CPA*(T-T0)/
(-DHrxn))
```

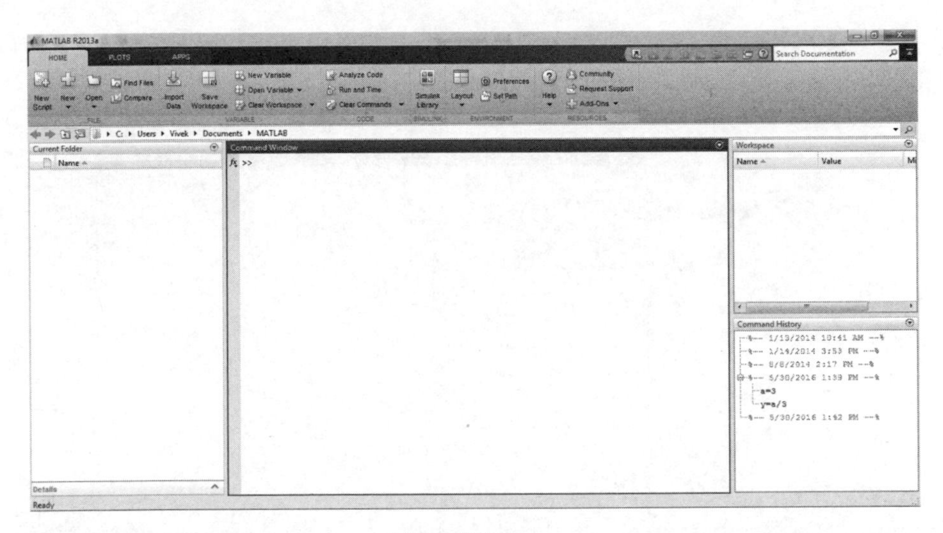

Figure A.3 Opening screen of MATLAB.

The variables correspond to the stated specifications. The command for the solve function involves equating the conversion from the material balance equation to the conversion from the energy balance. A double equal-to operator (==) is used in the solve function. MATLAB yields a solution of 537.4 K as the reactor temperature. The conversion at this temperature is 0.99, the same as that obtained using Mathcad and POLYMATH. The solution procedure and the solution are shown in Figure A.4.

The commands entered can be seen in the Command window. The window to the bottom right contains the command history, and the variables and their values can be seen in the window above it. The window to the left of the Command window contains the names of the files and folders.

Creating plots in MATLAB requires a slightly more involved procedure. First, the temperature variable is defined with a range of values. Then, two function files are created, which define and evaluate XMB and XEB as functions of the temperature. The details of defining a range for the temperature variable and creation of the two functions are not shown here and are left to the reader to explore. The plots of XMB (conversion from mass balance) and XEB (conversion from energy balance) as a function of temperature are then created by entering the following command:

PLOT(T, XMB, T, XEB)

This command creates the plot shown in Figure A.5.

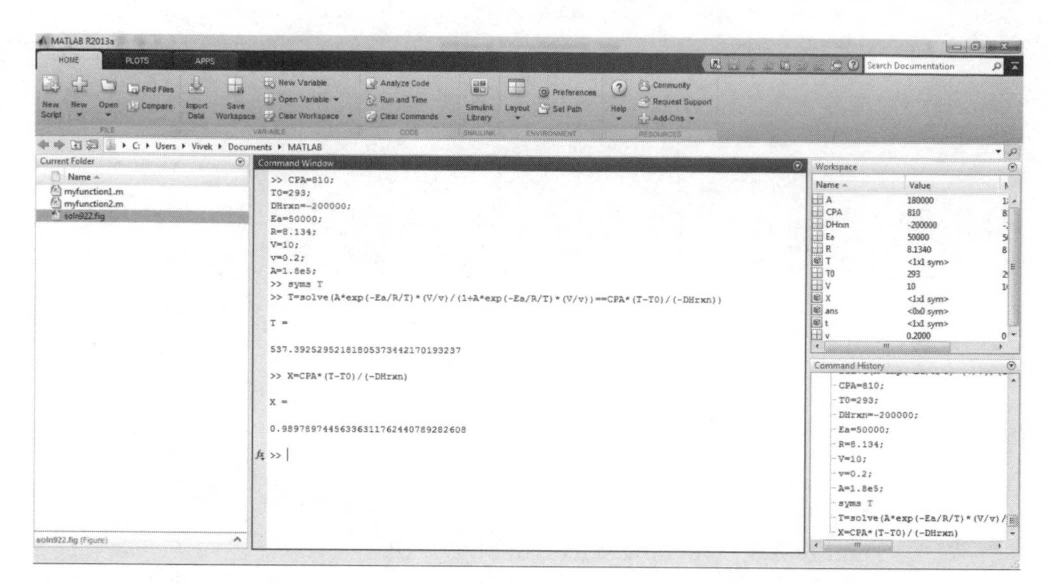

Figure A.4 MATLAB solution to example 9.2.2.

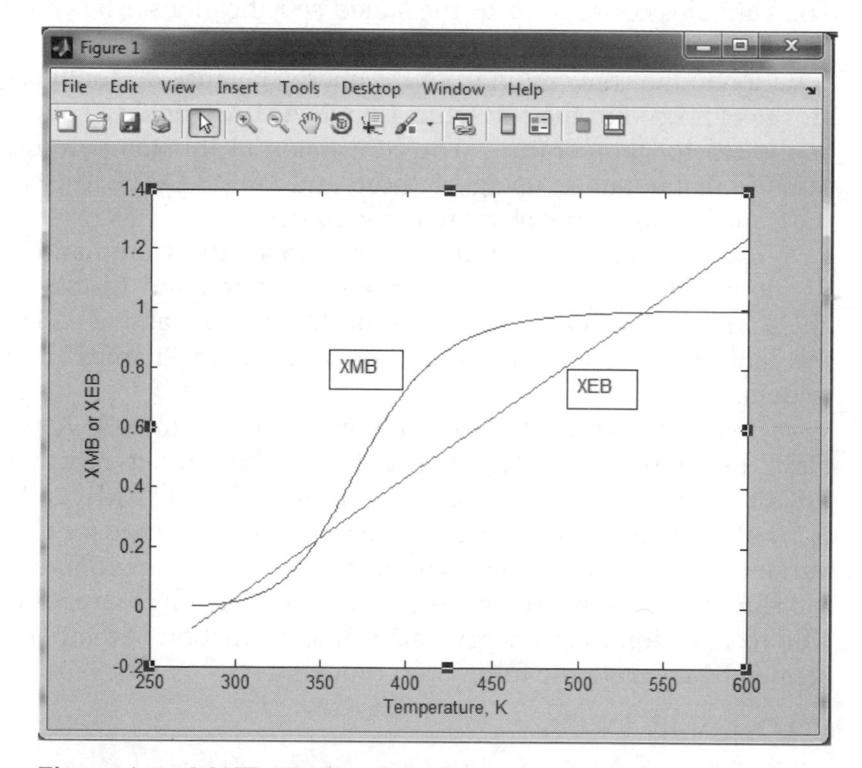

Figure A.5 MATLAB plot of conversions as a function of temperature.

It should be noted that alternative solution techniques for this example are possible in MATLAB. In any case, it can be seen that all three software programs (Mathcad, MATLAB, and POLYMATH) yield nearly identical solutions to the problem. Readers can verify for themselves that other programs (Excel, Mathematica, Maple, or any other program they employ for solving the problem) also yield the same results. The following section presents a brief comparison of the software packages.

A.2 Comparison of Software Packages

All the software packages offer some basic capability for computations, and then each one has its own unique features. An overview of similarities and differences is described in the context of various operations, such as basic arithmetic calculations, graphing, and programming capabilities.

A.2.1 Variables and Basic Operations

All the programs allow variables to be defined, and variable names are, in general, unrestricted. (Typically, computations in Excel reference the cell containing the variable; however, it is possible to assign a variable name.) All programs have built-in definitions of many variable names but also have provisions for overriding this definition. For example, MATLAB defines both i and j as $\sqrt{-1}$, but a user can assign a different value to them [2]. Similarly, Mathcad has built-in units and symbols: for example, by default, letters A, V, and L carry specific meanings of ampere, volt, and liter, respectively. However, the user can redefine these as variables to represent area, volume, length, or any other quantity [3]. Each program uses different syntax for assigning values to variables or defining the variable. For example, Mathcad uses a colon (:) as the assignment operator, whereas Maple uses a colon followed by equal to sign (:=) for assignment [3, 4]. Other programs use a simple equal-to sign (=) for assignment, whereas Mathcad uses it for evaluation and obtaining output. With practice, a user can gain expertise and use proper syntax. Most programs also provide color-coded errors and warnings if improper syntax or undefined variables are used in command or computation statements. Some programs have built-in units, such as SI units in the case of Mathcad, by default. For example, the default unit for pressure is *Pa* (pascals), and the values of any pressure variable will be output in this unit in Mathcad. Mathcad also has automatic conversion facilities for displaying the variable values in any user-chosen units. The user can simply click on the unit, change it to the desired unit (atm or psi, for example), and

Mathcad will automatically recalculate and display the proper number. Other programs require manual conversion of units.

All the programs use the same symbols (+, –, *, /) for arithmetic operations. Almost all programs allow a variable to be defined as a multidimensional array. In fact, most of the variables in MATLAB are considered as matrices (MATLAB stands for *Matrix Laboratory*) [2]. Most programs permit matrix computations and display error codes if invalid operations are attempted (addition of matrices of different dimension, etc.). All software programs have built-in functions for computing logarithms, trigonometric functions, and so on.

A.2.2 Equation Solving, Symbolic Computations, Plots, and Regression

All the programs have the capability to solve systems of linear algebraic equations as well as nonlinear transcendental equations. Excel offers this capability through its *Goal Seek* and *Solver* tools, and POLYMATH has its *LEQ* and *NLE* solvers. Solution techniques for Mathcad and MATLAB are presented in the book and in this appendix. Similar capability exists in Maple and Mathematica as well. Of course, each program may have multiple alternative pathways to arrive at the solution of the problem. A user, with practice, can master different techniques for solving problems. All programs also have the capability for solving many ordinary differential equations simultaneously. For example, POLYMATH has the *DEQ* program utility for solving differential equations, with initial conditions specified. Excel does not have a similar tool for solving differential equations; however, the grid structure of Excel makes it possible to manipulate the cells to adapt any numerical solution technique for ordinary as well as partial differential equations. Mathcad, MATLAB, Maple, and Mathematica also allow symbolic computations, such that users can obtain derivatives and integrals of expressions.

All programs have a graphing utility to generate plots—two-dimensional and, in some cases, three-dimensional. All programs have tools to manipulate the appearance of the graphs, including the color and shape of lines, markers, legends, axes scales, labels, and other elements. The ease of creation of plots and manipulating the format varies with the program, but users can easily learn to use the tools to enhance the effectiveness of the plots.

All programs also have regression tools for carrying out linear and nonlinear regression. The output of these tools generally includes statistical information about the goodness of fit and confidence in the estimates of the model parameters. Again, as with the plotting tools, users can acquire sufficient dexterity with practice.

A.2.3 Programming

Most of the computational problems encountered by a chemical engineer can be solved using any one of these programs (as well as some others not discussed in this appendix or the book). As mentioned earlier, the need to develop a customized program in a high-level language has been reduced considerably in recent times, primarily due to advances in the tools available in these software packages. However, a customized program may be needed for conducting computations in certain situations. Nearly all the software packages offer this programming capability, which is perhaps the biggest differentiating characteristic among these packages.

Excel offers a programming environment through its Excel VBA (Visual Basic–based) environment. Users can write the necessary programs (macros) and execute them for obtaining solutions. Although the programming capability in POLYMATH is limited, the POLYMATH outputs and statements can be converted and exported into MATLAB programs.

All other software packages offer the programming capability to varying degrees. All software packages have similar basic structure, with repeated calculations in loops, with *for*, *if*, and *while* statements serving as loop flow control tools. The versatility of programming is perhaps higher with MATLAB and Mathematica than with Maple and Mathcad.

Apart from these basic capabilities, each software package may contain specialized features that are suitable for many other applications. For example, Figure A.6 shows some of the tools available in MATLAB; the screen is brought up by clicking the APPS tab in the opening screen.

These apps provide MATLAB capabilities in signal processing, tuning of controllers, neural net fitting, optimization, and so on. These capabilities may be useful to a chemical engineer in some situations.

A.2.4 User Interface and Ease of Use

A fairly important consideration, particularly for a novice user, is the interface offered by the program and the ease of operation. The familiarity of Excel makes its interface one of the easiest to work with. The POLYMATH interface is equally friendly: it is not overloaded with icons and options, and it provides easy access to the particular computational problems (linear equations, differential equations, etc.). POLYMATH requires possibly the least amount of training, and a minimal time delay for a user to start wielding the computational tools. Of the remaining four, Maple and Mathcad have simpler, user-friendly interfaces, and MATLAB and Mathematica have comparable, more complex interfaces.

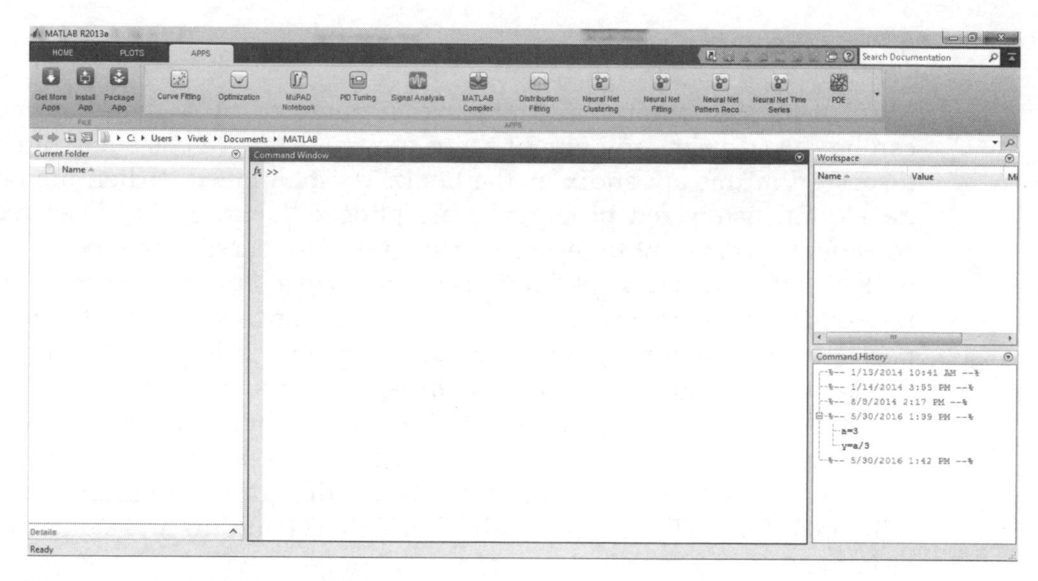

Figure A.6 APPS in MATLAB.

It should be noted that the programs with more complex interfaces also offer more capabilities with respect to programming and integration with other applications. An individual, with sufficient time and practice, can and will find the interface to be convenient. However, there is a significant learning curve for advanced software programs while the user gains familiarity with the syntax and command structure that is not intuitively clear [5]. The Help documentation offered by all of the programs is useful to a varying degree, with Excel and POLYMATH Help being the most useful, and MATLAB being less user friendly.

A.3 Summary

A large number of software programs ranging from the ubiquitous Excel to Mathematica and MATLAB are available in the market for performing engineering computations. In general, all the programs discussed in this appendix are effective in solving most of the computational problems encountered by a chemical engineer, such as the example problem of Chapter 9. A brief comparison of the capabilities of the various programs indicates that most of these programs offer similar computational capabilities. The comparison presented in this appendix is not intended to be comprehensive in nature, and many differences exist in the details of the features and executions. The software packages do differ in terms of their programming capabilities, and individuals who expect to or need to develop customized programs should explore these capabilities in detail before selecting a software package to procure.

References

1. Cutlip, M. B., J. J. Hwalek, H. E. Nuttall, M. Shacham, J. Brule, J. Widmann, T. Han, B. Finlayson, E. M. Rosen, and R. Taylor, "A Collection of 10 Numerical Problems in Chemical Engineering Solved by Various Mathematical Software Packages," *Computer Applications in Engineering Education*, Vol. 6, No. 3, 1998, pp. 169–180.
2. Constantinides, A., and N. Mostoufi, *Numerical Methods for Chemical Engineers with MATLAB Applications*, Prentice Hall, Upper Saddle River, New Jersey, 1999.
3. Adidharma, H., and V. Temyanko, *Mathcad for Chemical Engineers*, Second Edition, Trafford Publishing, Victoria, British Columbia, Canada, 2009.
4. White, R. E., and V. R. Subramanian, *Computational Methods in Chemical Engineering with Maple*, Springer-Verlag, Berlin Heidelberg, Germany, 2010.
5. Cutlip, M. B., and M. Shacham, *Problem Solving in Chemical Engineering with Numerical Methods*, Prentice Hall, Upper Saddle River, New Jersey, 1999.

APPENDIX B

Computations Using Process Simulation Software

Performing computations is an integral part of the responsibilities of a chemical engineer, particularly one engaged in the design of process units and plants. Chapters 4 through 9 provided an introduction to some of the problems and illustrated the solution techniques using two different computational tools. It can be understood at this point that a design engineer often faces computational problems that are significantly higher in complexity than those described in this book. For example, the flowsheet of the plant for ammonia synthesis was presented in Chapter 3, "Making of a Chemical Engineer." Although that flowsheet provided an overview of the complexity of the process, in reality, each unit of the process is considerably more complicated than it appears to be in that flowsheet. A chemical engineer must provide detailed information on each unit in the process, including all material and energy flows and their conditions (temperature, pressure, compositions, etc.) for a comprehensive description of the process plant. An ammonia synthesis plant is quite complex; however, even the simplest of the process plants will consist of several processing units, and material/energy balances over all the units need to be solved simultaneously. Further, very few of the chemicals in the process plants conform to the idealized behavior, whether present as a pure component or in a mixture. The incorporation of nonideality in the component and mixture behavior results in complicating the already complex relationships among the variables. At a fundamental computational level, a change from an ideal to a nonideal behavior transforms an explicit linear equation to a transcendental implicit equation, increasing the complexity of solution technique needed. Expressions that would be amenable to analytical integration in the case of ideal behavior will no longer remain so. Similarly, derivative forms increase in complexity, and solution techniques run the risk of failure or instability. It would seem that computational demands on a chemical engineer are a Herculean task, particularly since the engineer invariably operates under severe time constraints to deliver solutions.

Fortunately for the chemical engineer, there are several elegant computational tools available in the form of comprehensive process simulation software programs to perform the necessary calculations. These software programs have a built-in library of chemical compounds, quantitative descriptions of alternative thermodynamic models to describe nonideal behavior, and tools to draw and develop process flow diagrams and specify conditions and constraints. The process flow diagrams (or flowsheets) are drawn using built-in reaction, separation, heat transfer, fluid transport, and other process units. The software typically offers enough flexibility within the specifications of the process unit to permit specifying practically any operating condition. Most programs also allow user-defined, customized units as well as allow the user to define and include a chemical not present in the database and obtain comprehensive solutions for steady-state and dynamic simulations of chemical processes. Following are some of these software programs:

- *Aspen Plus*, Aspen Technology, Inc., Bedford, Massachusetts (www.aspentech.com)
- *CHEMCAD*, Chemstations, Inc., Houston, Texas (www.chemstations.com)
- *PRO-II*, Schneider Electric Software, Lake Forest, California (http://software.schneider-electric.com)
- *ProSimPlus*, ProSim S.A., France (www.prosim.net)

The representative capabilities of the process simulation software programs are illustrated through a simple separations problem solved using PRO-II.

B.1 Problem Statement: Adiabatic Flash Operation

An adiabatic flash is one of the simplest separation operations to effect a separation between components of a mixture based on the differences in volatility. A liquid stream is partially vaporized, with the result that the vapor produced is enriched in the more volatile components (lower boiling points) while leaving behind the liquid stream enriched in the less volatile (higher boiling temperatures). The adiabatic mode of operation for the separation means no heat is exchanged with the surroundings in the flash unit.

A process stream at 1000 psia and 200°F consisting of an equimolar mixture of acetylene and dimethyl formamide (DMF) enters a flash stage where the pressure is reduced to 200 psi. The total flow rate of the feed stream is 1 lbmol/h. The schematic representation of the process is shown in Figure B.1.

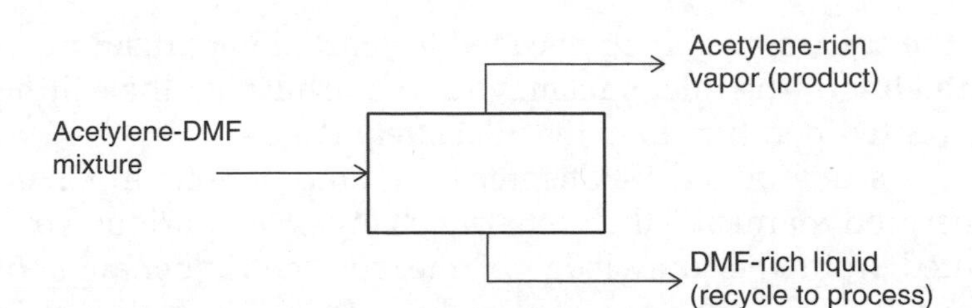

Figure B.1 An adiabatic flash stage.

A chemical engineer must conduct the material and energy balances on the process unit to obtain quantitative description of the unit. Following are the specific questions that need to be answered:

- What are the flow rates, temperature, and compositions of the product streams assuming that the volumetric behavior is described by Soave-Redlich-Kwong (SRK) equation of state (EOS)?
- What is the answer to the previous question if the behavior is instead governed by the ideal gas law?
- What is the percentage of error when ideality is assumed? Base the error calculations on results of the SRK behavior.

B.2 Theoretical Background and Solution Procedure

It is necessary to obtain a qualitative understanding of the process before attempting to obtain the solution to the problem. A brief theoretical background is presented in this section, followed by the solution procedure that illustrates the use of the process simulation software. A student will gain a detailed understanding of the process through the courses in chemical engineering thermodynamics and separation processes, as described in Chapter 3.

B.2.1 Theoretical Background

A flash operation is effected by reducing the pressure of a liquid stream in a vessel (flash drum) such that a fraction of the liquid fed vaporizes, with both the vapor and liquid product streams leaving in *equilibrium* with each other at the same pressure and temperature [1]. The more volatile component (acetylene in this example) is preferentially vaporized, resulting in a vapor stream that is enriched in acetylene and a liquid stream that is enriched in DMF, the less volatile component. An adiabatic flash operation is one in which no heat is exchanged with the surroundings. Since no heat is supplied to the stream, the energy needed for partial vaporization is derived from the enthalpy of the inlet stream, with the result that the exiting streams are at a lower temperature than the inlet stream.

As mentioned in Chapter 8, "Computations in Chemical Engineering Thermodynamics," the enthalpy changes for various streams can be computed from the knowledge of the volumetric behavior of the substances. Thus, an EOS that can accurately describe the volumetric behavior of the substances is an essential requirement for obtaining a solution to the preceding problem.

The SRK is a cubic EOS of the following form [2]:

$$P = \frac{RT}{v-b} - \frac{a \cdot \alpha(T,\omega)}{v(v+b)} \tag{B.1}$$

Here, v is the molar volume of the gas at pressure P and temperature T. R is the gas constant, a and b are characteristic constants for the substance, and $\alpha(T,\omega)$ is a complex function of the temperature T and acentric factor ω. Constants a and b are functions of the critical pressure P_C and critical temperature T_C. Each substance has its characteristic critical properties and acentric factor, the values of which are available from thermodynamic data sources. The characteristic parameters in the EOS are obtained from the following expressions:

$$a = 0.427\frac{R^2T_C^2}{P_C} \tag{B.2}$$

$$b = 0.08664\frac{RT_C}{P_C} \tag{B.3}$$

$$\alpha(T,\omega) = \left(1+\left(0.48+1.574\omega-0.176\omega^2\right)\left(1+T_r^{0.5}\right)\right)^2 \tag{B.4}$$

Here, T_r is the reduced temperature; that is, the temperature normalized by dividing it by the critical temperature ($T_r = T/T_C$).

The complexity of the EOS can be readily appreciated by contrasting it with the ideal gas law:

$$P = \frac{RT}{v} \tag{B.5}$$

The EOS and characteristic constants for the substance under consideration are available from the library and databases built into PRO-II. The solution of the problem involves creating the process unit and specifying the thermodynamic model governing the behavior of the system—the SRK EOS, or ideal gas law—as described in section B.2.2.

B.2.2 Solution Procedure

The first solution step involves creating a process flowsheet of the flash operation in PRO-II. Launching the software brings up the screen shown in Figure B.2.

Clicking OK, then selecting File, New, brings the home screen shown in Figure B.3 that allows creation of a new flow diagram. The user has the ability to bypass the window shown in Figure B.2 by changing the settings through the *Options* command from the menu bar. However, the window serves to provide a reminder about color coding used by PRO/II to convey messages/prompts about the input data and actions a user needs to take. A red box indicates that a user action is needed. A green box indicates default values of the data built in the software. The user may override these data by

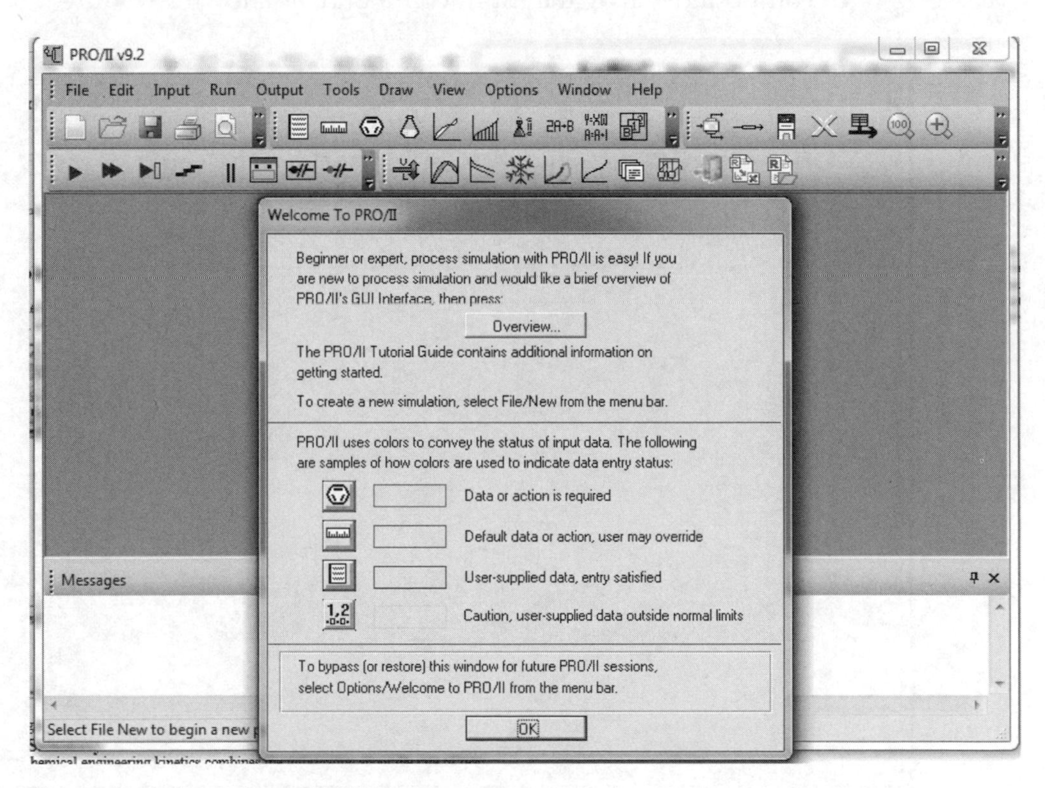

Figure B.2 Launching PRO-II software.

providing different values. If the user-supplied value is within the constraints of the variable, the box changes to the blue color. However, if the value is outside the normal range of the variable, the box color changes to yellow. In that case, the simulation may not run or may yield incorrect results if it does manage to run.

The open space is divided horizontally into two areas: the upper one is the workspace named *Flowsheet*, and the bottom one is named *Messages*, where warnings and other messages appear. The vertical panel on the right contains the templates for process equipment units and streams. The Streams tab is highlighted in Figure B.3, and the Flash Drum can be seen as the second process unit below Streams.

The following steps are followed to create the process flowsheet:

1. Add the flash unit to the flowsheet followed by the streams, as shown in Figure B.4. The red boxes indicate missing information and serve as prompts for user action.
2. Select the two components acetylene and DMF from the built-in library of compounds using the Input command, as shown in Figure B.5.

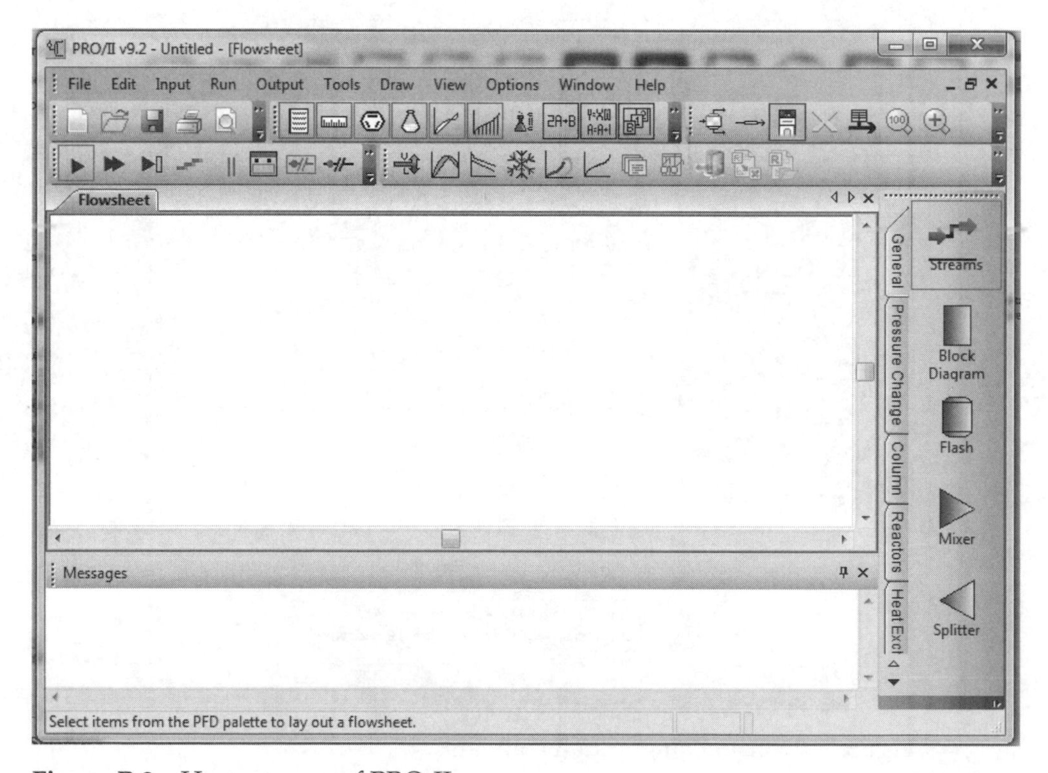

Figure B.3 Home screen of PRO-II.

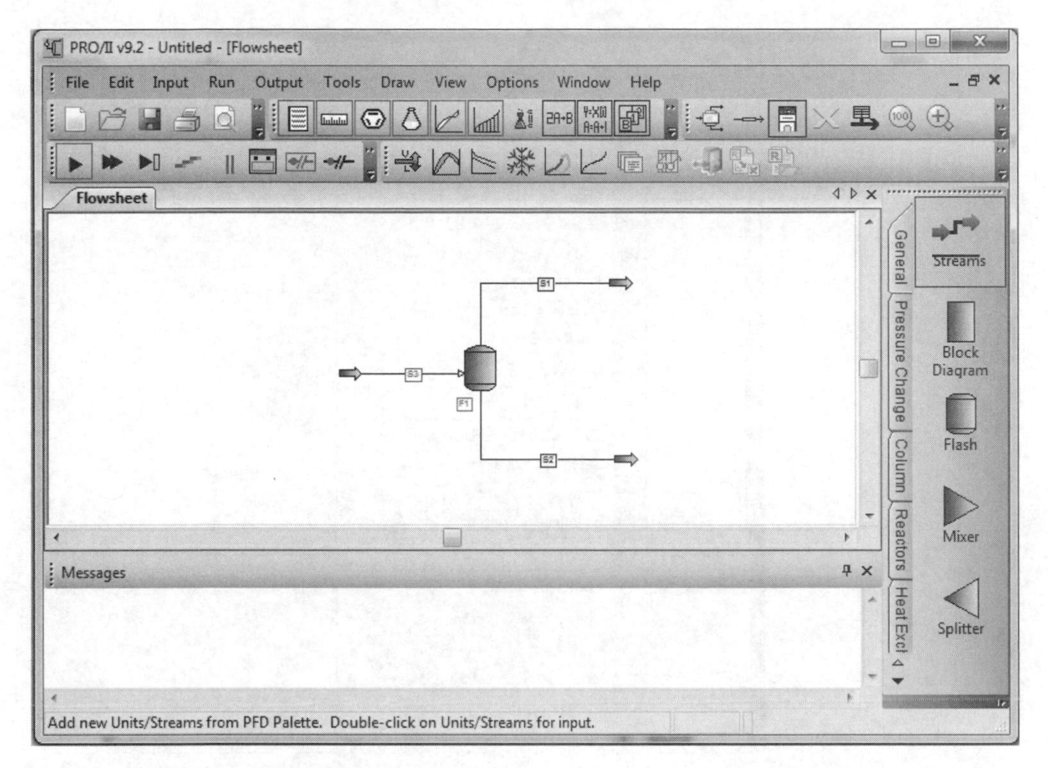

Figure B.4 Creating a flowsheet of the process.

3. The dropdown menu under the Input command contains the option for Thermodynamic Data, the second option below Component Selection. Click on Thermodynamic Data to bring up a dialog box, and select the SRK EOS, as shown in Figure B.6.
4. After the components and the thermodynamic models are specified, specify the inlet stream conditions by double-clicking on the inlet stream, which brings up the dialog box shown in Figure B.7.

   In general, various alternative specifications are possible for stream conditions. Based on the information provided, click the *Flowrate and Composition* button, enter the appropriate values, and specify the thermal condition of the stream using appropriate buttons. It should be noted that PRO-II has the flexibility to handle any system of units, with the Units of Measure selected using the dropdown menu under Input before specifying the stream conditions.
5. Double-click on the flash drum to bring up a dialog box where the mode of operation is specified, as shown in Figure B.8.

   The flash operation requires two specifications: the first one comes from the specified pressure (200 psia), and the second one is zero

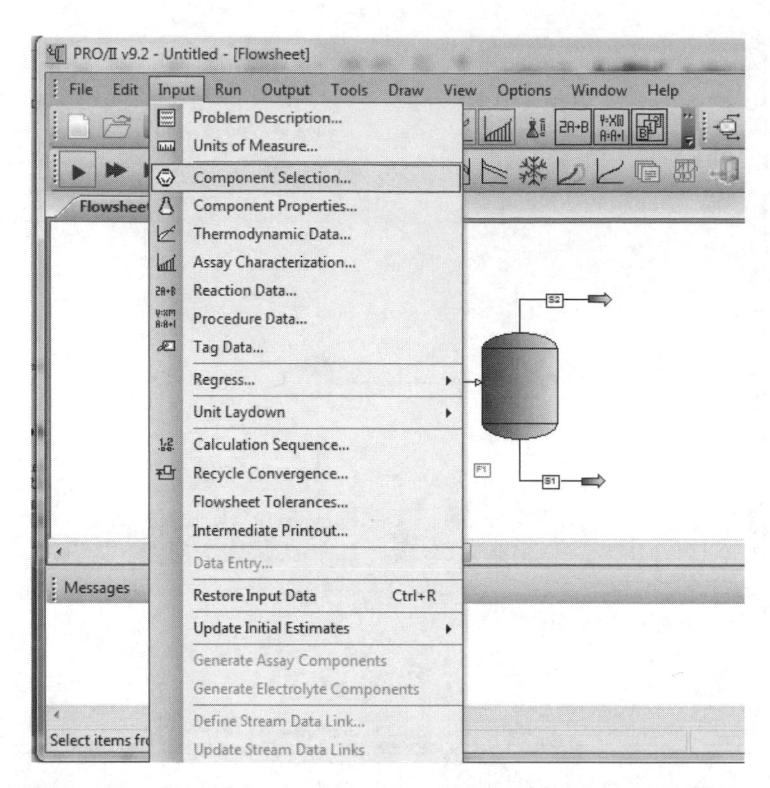

Figure B.5 Specifying components for the problem in PRO-II.

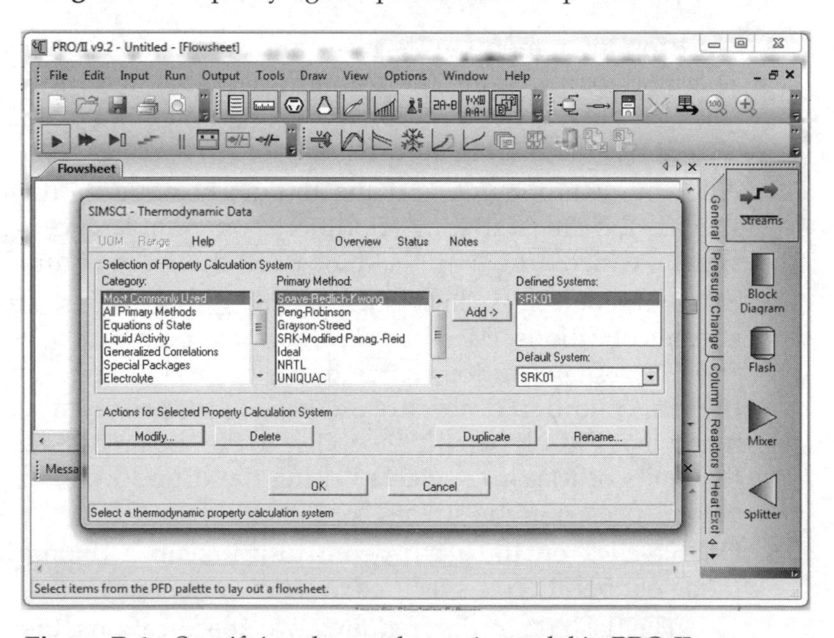

Figure B.6 Specifying thermodynamic model in PRO-II.

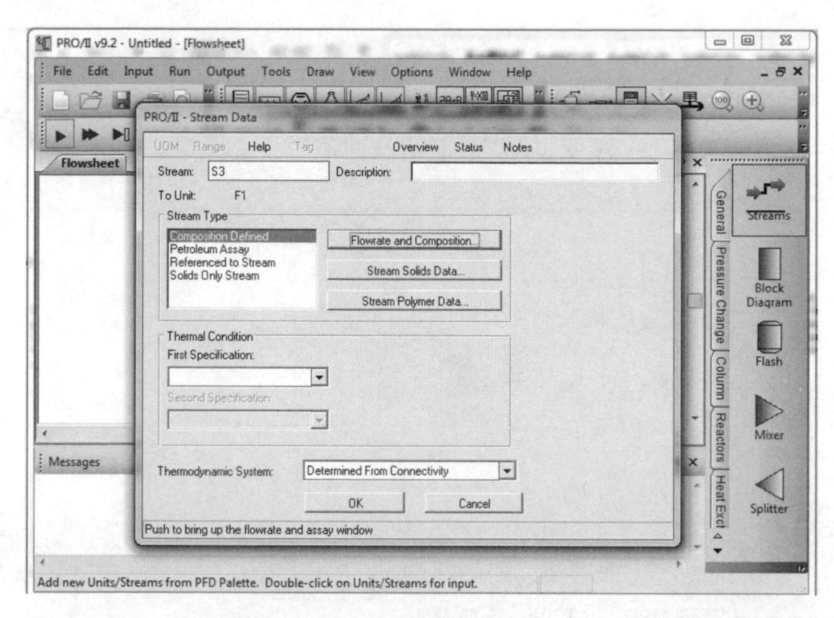

Figure B.7 Specifying stream properties and conditions in PRO-II.

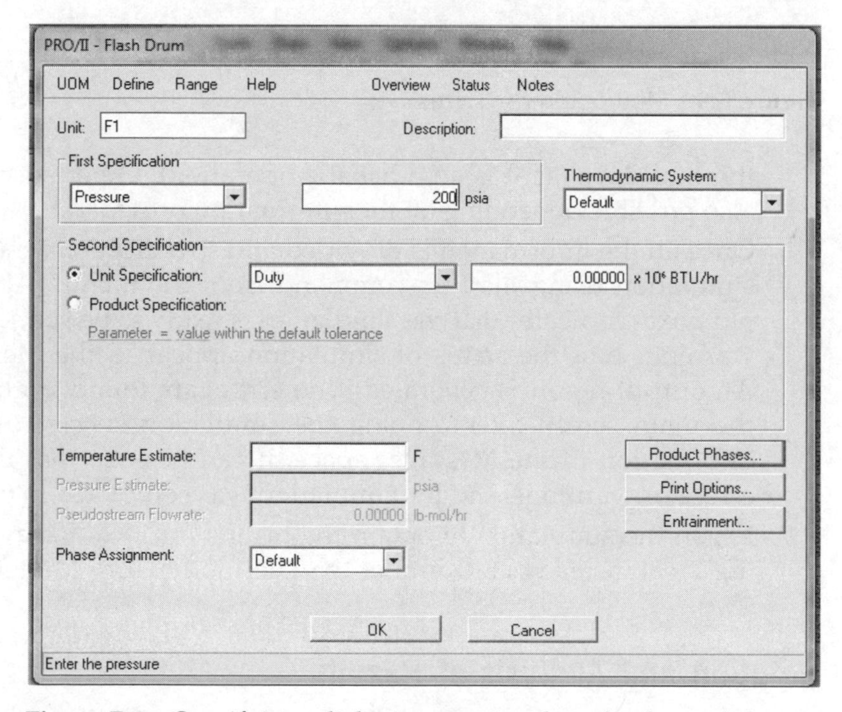

Figure B.8 Specifying adiabatic operational mode: thermal duty set to 0.

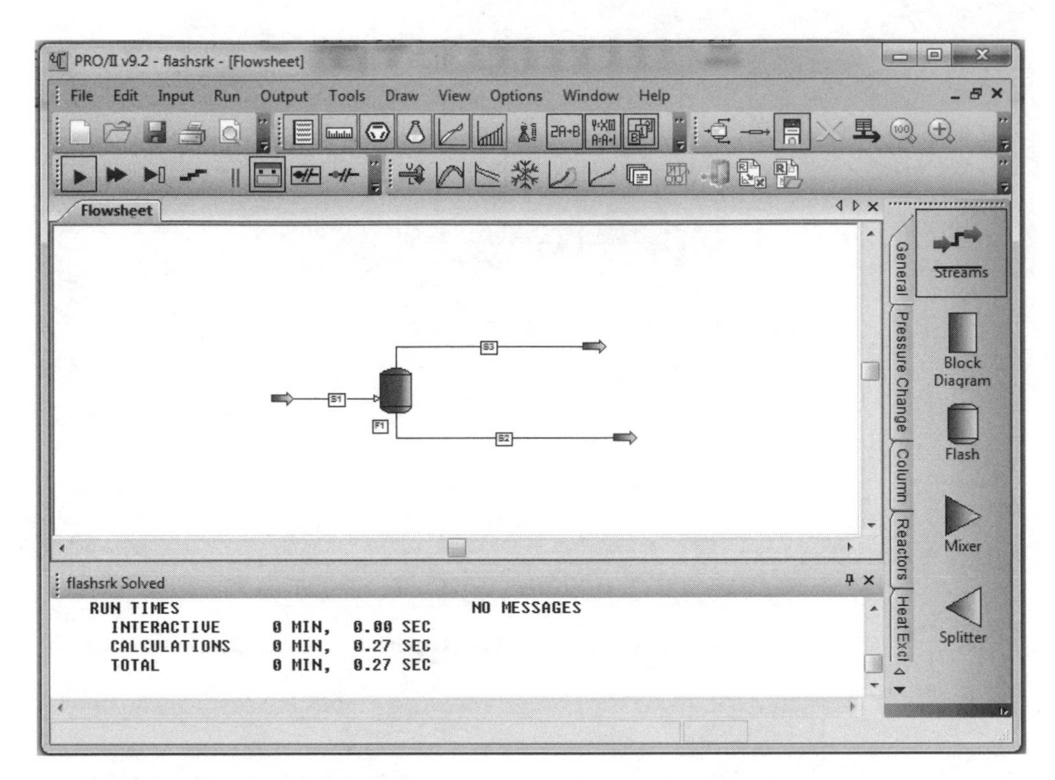

Figure B.9 Simulation run status.

thermal duty (as it is an adiabatic operation). Other specifications are also possible depending on the information provided.

6. Once all the information is provided and specifications entered, run the simulation using the Run command from the menu. (Absence of any red boxes indicates that the simulation is ready to be run.) Any errors or warnings and the status of simulation appear in the Messages space. An output report is generated if no errors are found. Figure B.9 shows the main screen after running the simulation successfully. From the information in the Messages space, it can be seen that there were no errors or warnings, and the simulation was completed in 0.27 seconds.
7. Rerun the simulation to obtain the results for ideal behavior by selecting Ideal instead of SRK under the Thermodynamic Model.

B.3 Solution and Analysis of Results

The resulting solution using the SRK model is presented first, followed by the comparison with results obtained on the basis of ideality.

B.3.1 Simulation Using SRK Model

The results of the simulation in terms of the stream temperatures and flow rates using SRK EOS are shown in Figure B.10.

It can be seen from the figure that ~41% of the feed exits as the vapor product that is essentially entirely acetylene. The mole fraction of DMF in the vapor is 0.0037, which although quite low, may or may not be an acceptable impurity level for the acetylene product depending on its intended application. If a further reduction in DMF content is desired, other separation operations, such as absorption or adsorption, may be used. The remaining 59% of the feed exits as a liquid, which is mostly DMF. However, it contains significantly higher levels of acetylene (mole fraction 0.1565), which would probably require further processing for separation before the DMF is recycled to the process for use as solvent.

B.3.2 Comparison of Results with Ideal Behavior

As mentioned, the simulation was also run with the ideal gas law serving as the thermodynamic model. The comparison of results for the ideal and nonideal cases is shown in Table B.1.

As can be seen from the table, the ideality assumption yields a temperature of ~197°F, while the actual temperature will be ~142°F. This difference in temperature has a substantial impact on the energy consumption of the process. If the streams are to be raised back to the feed temperature, the ideality assumption will predict a much lower energy requirement than what will actually be needed. Although both the models indicate that the vapor

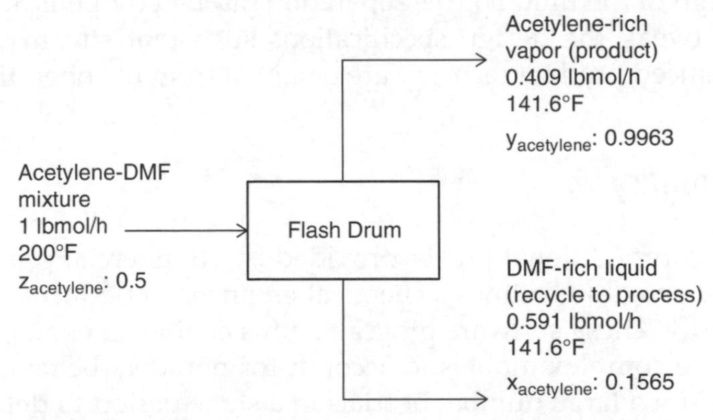

Figure B.10 Simulation results for adiabatic flash of DMF-acetylene mixture.

Table B.1 Comparison of Results—Effect of Mixture Nonideality

		Ideal		SRK	
		Vapor	Liquid	Vapor	Liquid
Temperature (°F)		197.25		141.64	
Flow Rates, lbmol/h		0.463	0.537	0.409	0.591
Mole Fractions	AC	0.9902	0.0765	0.9963	0.1565
	DMF	0.0098	0.9235	0.0037	0.8435

stream will be at least 99% acetylene, the purity level for acetylene in the ideal case may not be sufficient. The flow rates of the product streams differ significantly. The ideal assumption overpredicts the vapor flow rate by ~13%, meaning there will be substantially lower quantity of acetylene product in reality. Further, the DMF being recycled to the process will be lower in purity—the ideality assumption predicts a liquid stream 92% pure in DMF, but the actual purity is only ~84%. In reality, a significant quantity of acetylene is being recycled to the process through the liquid stream, which has substantial impact on the process economics through not only a lower product output but also additional expense of reprocessing.

The need to have as realistic a model of the process as possible and use of appropriate thermodynamic model should be abundantly clear to an engineer from these results. The value of the process simulation software should also be understood and appreciated by the engineer. Not only is the program able to incorporate complex thermodynamic models for system description, it also provides an extremely rapid solution. The program provides the design engineer with a powerful tool to vary the process conditions and simulate a large number of scenarios in very little time in order to optimize the design of the unit. For the separation discussed in this appendix, an engineer can tweak the design specifications such that streams of higher purity or enhanced product recovery are obtained from the operation.

B.4 Summary

The computational power provided by commercial process simulation software programs allows a chemical engineer to perform complex calculations rapidly. These software programs thus enable an engineer to not only incorporate complex models to account for nonideal behavior of substances but also run a large number of trials in a short period to determine the effects of various parameters.

References

1. Seader, J. D., E. J. Henley, and D. K. Roper, *Separation Process Principles: Chemical and Biochemical Operations,* Third Edition, John Wiley and Sons, New York, 2010.
2. Fink, J. K., *Physical Chemistry in Depth,* Springer-Verlag, Berlin Heidelberg, Germany, 2009.

INDEX

B

C

D

G

H

I

S

T

U

V

W

X

Z